ISSN 0323-3898

Volume 11 · 1980 · Number 4

MATHEMATISCHE OPERATIONSFORSCHUNG UND STATISTIK

# optimization

OPERATIONS RESEARCH AND MATHEMATICAL STATISTICS

ИССЛЕДОВАНИЕ ОПЕРАЦИИ И МАТЕМАТИЧЕСКАЯ СТАТИСТИКА

RECHERCHE OPERATIONELLE ET STATISTIQUE MATHEMATIQUE

4

Akademie-Verlag · Berlin

EVP 20,– M 32 585

## Instructions for Authors

**Submission of papers.** Papers to be submitted for publication or books to be reviewed should be sent to

PROFESSOR K.-H. ELSTER
Chief Editor, Optimization
TH Ilmenau
Sektion Mathematik, Rechentechnik und ökonomische Kybernetik
DDR - 6300 Ilmenau, Ehrenberg, Block G

Papers may be written in English, French, German or Russian. It is assumed, that the papers have not been published previously. The original (or xerox copy) should be submitted with two additional copies, which will not be returned to the author. Authors are held fully responsible for all information included in their papers.

**Preparation of manuscripts.** Manuscripts should be typewritten, at least double-spaced with wide margins at all four sides. Before having a paper typed for submission, please examine a recent issue of Optimization carefully and follow the arrangement of formulae, references, tables etc. Headings should be numbered by arabic numbers. The exact position of mathematical characters must be absolutely unambiguous. Distinguish between "oh" and "zero", "el" and "one", "kappa" and "kei", etc.

Text (with exception of mathematical symbols and formulae) may be stressed manually in the following ways:

| | |
|---|---|
| Bold | straight black underline |
| Italics | waved blue underline |
| Spaced types | dotted underline |
| Small capitals | waved red underline |

The following script annotation rules shall be observed for mathematical symbols and formulae:

Greek letters (excluding summation and product signs) should

| | | |
|---|---|---|
| be underlined | | with a red line |
| bold roman types | (**A**, **B**, . . .) | with two black lines |
| bold italics | (***A***, ***B***, . . .) | with two blue lines |
| gothic types | ($\mathfrak{A}$, $\mathfrak{B}$, . . .) | with a green line |
| bold gothic types | ($\boldsymbol{\mathfrak{A}}$, $\boldsymbol{\mathfrak{B}}$, . . .) | with two green lines |
| straight types | (A, B, . . .) | with a violet line |
| script | ($\mathcal{A}$, $\mathcal{B}$, . . .) | with a brown line |

The author may list on a separate sheet entitled "Hints for the printer" information which will help to avoid misinterpretations.

**Summary and key words.** A selfcontained summary, if possible in three of the languages mentioned, should be typed on a separate page. Its length should not exceed 150 words. It should be followed by a list of not more than 10 key words indicating the contents of the paper.

**Tables and illustrations.** Each table and each illustration should be prepared on a separate sheet. They should be numbered consecutively throughout the paper and be referred to in the text as Table 1, Fig. 1 etc. Their approximate position for insertion should be indicated in the margin of the manuscript. The number of illustrations should be kept as low as possible.

**Footnotes.** Should be reduced to minimum and, where possible, should be replaced by remarks in the text. The first footnote should be the present addresses of the authors on the first page.

**Proofs and reprints.** Authors will receive page proofs (in two sets). One corrected set of proofs should be returned to the Editor within on week. Authors will be charged for alterations in excess 10% of the cost of composition.

75 reprints are offered free; up two 250 additional reprints are provided at cost price.

ISSN 0323-3898

# optimization

MATHEMATISCHE OPERATIONSFORSCHUNG UND STATISTIK

SERIES OPTIMIZATION

Volume 11
1980
Number 4

Published at
Sektion Mathematik, Rechentechnik
und ökonomische Kybernetik
Technische Hochschule Ilmenau



AKADEMIE-VERLAG · BERLIN

**Editorial Policy**

The journal publishes theoretical and applied papers related to the fields of linear nonlinear, stochastic, parametric, discrete and dynamic programming, control theory, game theory, mathematical modelling of economical and technological processes, inventory, queueing theory, reliability, renewal theory, network methods etc. It is expected, that the papers give interesting and novel contributions to the theory at a high mathematical level or a stimulating presentation of ideas in foundations or applications. All papers should begin with a clear introduction including practical or theoretical consequences and connections to previous works. The trend to highest generality should not obscure the essential ideas.
A special section will be devoted to survey papers on theory and methods in interesting areas of mathematical programming and optimization techniques. Furthermore, the journal publishes proceedings of conferences, book reviews and announcements on related topics. Experts will be asked to give an opinion on each paper submitted for publication.

Math. Operationsforsch. Statist., Ser. Optimization, Vol. 11 (1980) No. 4, 519–530

# Duality for Quasi-Concave Programs with Explicit Constraints

T. R. JEFFERSON[1] and C. H. SCOTT[1]

**Summary:** The idea of duality is now well established in the theory of concave programming. The basis of this duality is the concave conjugate transform. This has been exemplified in the development of generalised geometric programming. Much of the current research in duality theory is focused on relaxing the requirement of concavity. Here we develop a duality theory for mathematical programs with a quasiconcave objective function and explicit quasiconcave constraints. Generalisations of the concave conjugate transform are introduced which pair quasiconcave functions as the concave conjugate transform does for concave functions. Optimality conditions are derived relating the primal quasiconcave program to its dual. This duality theory was motivated by and has implications in certain problems of mathematical economics. An application to economics is given.

## 1. Introduction

An essential concept in the analysis of mathematical programs is the idea of duality. This has furnished new approaches and interpretational insights and provided the basis for many powerful algorithms. To date, most of this theory has concentrated on convex mathematical programs whereas the vast spectrum of nonconvex programs has remained relatively untouched.

Successful approaches to duality for convex programs have generally been along *Lagrangian* lines [1] or used conjugate function theory [2]. The latter approach, suitably extended by ROCKAFELLAR [3] and PETERSON [4] has had substantial impact on current theoretical developments. However the assumption of convex (concave) functions is too restrictive for some applications in such areas as economics. One often deals with utility functions which are more appropriately of quasiconcave form rather than concave.

A pioneering analysis of quasiconcave programs in $R^n$ was given by ARROW and ENTHOVEN [5] and some duality results along *Lagrangian* lines have been given by MANGASARIAN [1]. Here we follow the definition of quasiconcavity found in MANGASARIAN [1].

In an earlier paper [6], we developed a duality theory for mathematical programs with a strictly concave objective function. The novelty of this approach is

[1] University of New South Wales, Kensington, N.S.W. 2033 P.O.Box 1, Australia.

the use of the utility transform in much the same way that one uses a conjugate transform for convex mathematical programs.

The utility transform and the duality based on it is similar to the pioneering work of GREENBERG and PIERSKALLA [7] and further developed by CROUZEIX [8]. The difference is that our method seeks to separate the objective function from the cone condition and explicit constraints in the calculation of the dual. This simplifies the calculaton of the dual mathematical program and makes the theory more amenable to application. Details can be found in the following section.

In this paper we extend the theory developed in [6] to handle explicit quasi-concave constraints. Such an extension necessitates the introduction of a further transform: the profit transform. This work leads us to believe that the development of a successful duality theory rests with the choice of the appropriate transformation.

In the following section we present the properties of the utility, profit and conjugate transforms. In the third section, we use the utility and profit transforms to develop a duality theory for quasi-concave programs with explicit constraints. This theory is illustrated by an example from economics in the final section.

## 2. Transforms

The choice of transform(s) is the key to a duality theory. The duality theories of PETERSON [4], and ROCKAFELLAR [3] are based on the conjugate transform.

**Definition:** The *concave conjugate transform* of a concave function $f$ defined on a convex set $F$ is the pair $[g, G]$ defined by

$$g(y) = \inf_{x \in F} [\langle x, y\rangle - f(x)] \tag{1}$$

and

$$G = \{y \mid \inf_{x \in F} [\langle x, y\rangle - f(x)] > -\infty\}. \tag{2}$$

For $x \in F$, $y \in G$ we have the *conjugate inequality*

$$f(x) + g(y) \leqq \langle x, y\rangle. \tag{3}$$

**Definition:** The *supergradient* of a concave function $f(x)$, $x \in F$ is the set $\partial f(x)$ defined by

$$\partial f(x) = \{y \mid f(z) - f(x) \leqq \langle z - x, y\rangle,\ \forall z \in F\}.$$

Inequality (3) holds with equality iff

$$y \in \partial f(x) \quad \text{and} \quad x \in \partial g(y).$$

**Definition:** The *hypograph* of a pair $[f, F]$ of a function $f$ defined on a set $F$ is the set

$$\{(x, \xi) \mid x \in F \text{ and } \xi \leqq f(x)\}.$$

**Definition:** A concave or quasiconcave function is said to be *closed* if its hypograph is a closed set.

The conjugate transform guarantees $g$ defined in (1) to be closed and concave (which implies continuity) whether $f$ is concave or not. This is quite a strong result. Moreover PETERSON has been able to develop a theory which handles explicit concave constraints. An explicit constraint $f(x) \geqq 0$, $x \in F$ is paired with a function $g^+$ defined on a set $G^+$ where

$$g^+(y, \nu) = \begin{cases} \nu g(y/\nu), & \nu > 0 \\ \inf_{x \in F} \langle x, y \rangle, & \nu = 0 \end{cases} \tag{4}$$

and

$$G^+ = \{(y, \nu) \mid \nu > 0,\ y/\nu \in G\} \cup \{(y, \nu) \mid \nu = 0,\ \inf_{x \in F} \langle y, x \rangle > -\infty\}$$

$[g, G]$ is defined by (1), (2).

$g^+$ is concave and positively homogeneous of degree one by construction. $[g^+, G^+]$ is called the positive homogeneous extension of $[g, G]$. The conjugate inequality here is

$$g^+(y, \nu) \leqq \langle x, y \rangle \tag{5}$$

for $x \in F$, $(y, \nu) \in G^+$ and $f(x) \geqq 0$.
Inequality (5) holds at equality iff

$$y \in \nu \partial f(x) \quad \text{or} \quad x \in \partial g(y/\nu), \quad \nu > 0$$

or

$$x \in \{z \mid \langle y, z \rangle = \inf_{z' \in F} \langle y, z' \rangle\} \quad \text{for} \quad \nu = 0.$$

In order to handle quasiconcave functions a weaker transform is required. In [6], we developed the utility transform. This transform generates a quasiconcave functon which has meaning in economics and game theory.

**Definition:** The *utility transform* of a quasiconcave function $u$ defined on a convex set $U$ is the pair $[v, V]$ defined by

$$v(p) = \inf_{x \in U} \{-u(x) \mid \langle p, x \rangle \leqq 0\}$$

and

$$V = \{p \mid \inf_{x \in U} [-u(x) \mid \langle p, x \rangle \leqq 0] > -\infty\}.$$

The utility transform of $u$ is quasiconcave and positively homogeneous of degree zero regardless of the assumptions on $u$. For $x \in U$, $p \in V$, $\langle p, x \rangle \leqq 0$ we have the utility inequality

$$u(x) + v(p) \leqq 0. \tag{6}$$

**Definition:** The *local supergradient* of a quasiconcave function $u(x)$, $x \in U$ is a set $\partial^{\text{loc}} u(x)$ defined by

$$\partial^{\text{loc}} u(x) = \Big\{p \mid \lim_{\alpha \to 0} \frac{u\,(x + \alpha \Delta x) - u(x)}{\alpha} \leqq \langle \Delta x, p \rangle,$$
$$\forall \Delta x \text{ s.t. } x + \beta \Delta x \in U \text{ for some } \beta > 0\Big\}.$$

Provided $[u, U]$ is strictly quasiconcave and closed we have equality in (6) iff $\lambda p \in \partial^{\mathrm{loc}} u(x)$, $\lambda > 0$. See [6] for details.

In [7], GREENBERG and PIERSKALLA introduced the *z-quasi-conjugate* of a quasiconcave function $u(x)$ defined by

$$u_z^*(p) = z + \inf\{-u(x) \mid \langle x, p\rangle \leqq z\}.$$

However the $z$ variable is difficult to handle in the dual. CROUZEIX [8] has used the following variation to improve the transform

$$q_{x_0}(p) = \inf\{-u(x) \mid \langle x - x_0, p\rangle \leqq 0\}.$$

However to separate the transform from the cone condition we require the inequality $\langle x - x_0, p\rangle \leqq 0$ to be a cone. This is the case when $x_0 = 0$ or $z = 0$ for the $z$-quasi-conjugate. With this restriction both transforms reduce down to the utility transform.

To handle explicit quasiconcave constraints we use the profit transform.

**Definition:** The *profit transform* of a quasiconcave function $f$ defined on a convex set $F$ is the pair $[h, H]$ defined by

$$h(p) = \inf_{x \in F} (\langle p, x\rangle \mid f(x) \geqq 0) \tag{7}$$

and

$$H = \{p \mid \inf_{x \in F} (\langle p, x\rangle \mid f(x) \geqq 0) > -\infty\}.$$

**Theorem 1:** *Let $[h, H]$ be the profit transform of a quasiconcave function $f$ defined on a convex set $F$. $h$ is concave and positively homogeneous of degree one. $H$ is a convex cone.*

The proof follows directly from the definitions.

**Lemma 1:** *$[h, H]$ is closed.*

Proof: The result is proved by observing that $[h, H]$ is the concave conjugate transform of the zero function defined on $F \cap \{x \mid f(x) \geqq 0\}$. The concave conjugate transform is closed. See [3]. ∎

By definition of the profit transform we have the profit inequality holding for $x \in F$, $f(x) \geqq 0$ and $p \in H$:

$$h(p) \leqq \langle p, x\rangle. \tag{8}$$

**Theorem 2:** *Let $f(x)$ be upper semicontinuous and $F$ closed.*

(i) (a) *the profit inequality* (8) *holds at equality iff*

$$x \in \partial h(p) \quad or$$

(b) *$p$ belongs to the supergradient of $[0; f(x) \geqq 0, x \in F]$ at $x$.*

(ii) *the latter condition* (b) *can be refined to $p \in \lambda \partial^{\mathrm{loc}} f(x)$, $\lambda \geqq 0$.*

Proof: (i) is an immediate consequence of the fact that $[0; f(x) \geqq 0, x \in F]$ is a closed concave function and the properties of the conjugate transform.

(ii) is obtained by considering the minimization problem:

$$h(p) = \min_{x \in F} \{\langle p, x\rangle \mid f(x) \geqq 0\}. \quad ∎$$

The necessary and sufficient conditions for $x$ to be the optimal solution are

$$p \in \lambda \partial^{loc} f(x), \quad \lambda \geqq 0$$
$$f(x) \geqq 0, \qquad x \in F.$$

The profit transform is related to other transforms appearing in the literature. If the condition of quasiconcavity is strengthened to concavity we have

$$h(p) = \sup_{v} \{g^+(p, v) \mid (p, v) \in G^+\}$$

where $h(p)$ is defined in (7) and $[g^+, G^+]$ in (4). Thus when the functions are concave the profit transform yields the same function as does the concave conjugate transform theory.

The motivation of the name profit transform comes from economics. If $f(x)$ is a production function, $-h(p)$ is known in the literature as the profit function. Further details are given in [9].

The utility transform described earlier and the profit transform can be thought of as a pair. The profit transform can be written as

$$\inf_{x \in F} \{\langle p, x \rangle \mid -f(x) \leqq 0\}.$$

Interchanging the inner product with the negative of a concave function produces the utility transform of $f(x)$.

**Lemma 2:** *Let $f(x)$ be positively homogeneous of degree one and upper semi-continuous. Let $F$ be a closed convex cone. Then the profit transform $h(p)$ is identically zero on $H$.*

Proof: This is a direct consequence of the definition of the profit transform. ■

In the following sections we shall assume that the functions handled by the profit transform satisfy the assumptions of lemma 2. These are production functions with constant returns to scale [9], [10].

## 3. Duality Using the Utility and Profit Transforms

Consider the following primal quasiconcave program.

Program A: Maximize $u(x^0)$
subject to

$$f_i(x^i) \geqq 0, \quad i = 1, 2, \ldots, m$$
$$x^0 \in U$$
$$x^i \in F_i, \qquad i = 1, 2, \ldots, m$$
$$x = (x^0, x^1, \ldots, x^m) \in X$$

where

$u$ is a closed strictly quasiconcave function defined on $U$. $f_i$, $i = 1, \ldots, m$ are closed quasiconcave functions which are positively homogeneous of degree one defined on $F_i$, $i = 1, \ldots, m$. $X$ is a convex cone.

The dual of program A is:
Program B: Maximize $v(p^0)$
subject to

$$p^0 \in V$$
$$p^i \in H_i \quad i=1, 2, \ldots, m$$
$$p=(p^0, p^1, \ldots, p^m) \in \Pi$$

where $[v, V]$ is the utility transform of $[u, U]$. $[0, H_i]$ is the profit transform of $[f_i, F_i]$, $i=1, \ldots, m$. $\Pi$ is the dual cone to $X$ defined by

$$\Pi=\{p \mid \langle p, x\rangle \leqq 0, \ \forall x \in X\}.$$

At optimality, the primal and dual solutions satisfy the following conditions:

$$x \in X, \quad x^i \in F_i, \ f_i(x^i) \geqq 0, \quad i=1, 2, \ldots, m \tag{9}$$
$$p \in \Pi, \quad p^i \in H_i, \quad i=1, 2, \ldots, m \tag{10}$$
$$\langle x, p\rangle = 0 \tag{11}$$
$$u(x^0)+v(p^0)=0 \tag{12}$$
$$\lambda p^0 \in \partial^{\mathrm{loc}} u(x^0), \quad \lambda > 0 \tag{13}$$
$$p^i \in \nu_i \partial^{\mathrm{loc}} f_i(x^i), \quad \nu_i \geqq 0, \quad i=1, 2, \ldots, m. \tag{14}$$

In the following theorem we will relate the optimality conditions (9—14) with optimal solutions for programs A and B.

**Theorem 4:** *The optimality conditions* (9—14) *are necessary and sufficient for optimality of programs* A *and* B.

Proof: Let $\bar{x}$ be an optimum for program A. $\bar{x}$ must be feasible and therefore satisfy (9). Since $\bar{x}$ is optimal consider a local supergradient $p$ such that

$$p^i \in \nu_i \partial^{\mathrm{loc}} f_i(x^i), \quad i=1, \ldots, m$$

and

$$\lambda p^0 \in \partial^{\mathrm{loc}} u(x^0).$$

Consider any change $\Delta x$ from $\bar{x}$ such that $\bar{x}+\Delta x \in X$. Thus $\langle \Delta x, p\rangle \leqq 0$. In particular consider $\Delta x = \bar{x}$ or $-\bar{x}$ and $\langle p, x\rangle = 0$. Furthermore $p \in \Pi$. (12) is by the utility and profit inequalities.

Suppose $\bar{p}$ is an optimum for program B. $\bar{p}$ must therefore satisfy (10). By construction there exist $\lambda$, $x^0$, $\nu_i$, $x^i$, $i=1, 2, \ldots, m$ s.t.

$$\lambda \bar{p}^0 \in \partial^{\mathrm{loc}} u(x^0), \quad \lambda \geqq 0$$
$$\bar{p}^i \in \nu_i \partial^{\mathrm{loc}} f_i(x^i), \quad \nu_i \geqq 0, \quad i=1, 2, \ldots, m$$

which are conditions (13) and (14).

Thus

$$u(x^0)+v(\bar{p}^0)=\langle \bar{p}^0, x^0\rangle$$

and by lemma 2 and the definition of the profit transform

$$\langle \bar{p}^i, x^i\rangle = 0, \quad i=1, \ldots, m.$$

Summing

$$u(x^0)+v(\bar{p}^0)=\langle \bar{p}, x\rangle.$$

For $\bar{p}$ to be optimal, any feasible direction $\Delta p$ must satisfy $\langle x, \Delta p\rangle \leqq 0$, $\bar{p}+\Delta p \in \Pi$. In particular it must hold for $\bar{p}$ and $-\bar{p}$. Therefore $\langle x, \bar{p}\rangle = 0$ yielding (11) and (12).

Assume now that the optimality conditions (9—14) hold for $\bar{x}$, $\bar{p}$. Suppose there exists an $\hat{x}$ satisfying (9) such that $u(\hat{x}^0) > u(\bar{x}^0)$ then either

(i) $\langle \bar{p}^0, \hat{x}^0\rangle \leqq 0$ and $u(\hat{x}^0)+v(\bar{p}^0) > 0$ which is impossible or
(ii) $\langle \bar{p}^0, \hat{x}^0\rangle > 0$ which contradicts the feasibility of $\bar{p}^0$ or $\hat{x}^0$ by Theorem 1 of [6].

Suppose now that $\bar{p}$ is not optimal and there exists a $\hat{p}$ such that

$$v(\hat{p}^0) > v(\bar{p}^0).$$

If $\langle \hat{p}^0, \bar{x}^0\rangle \leqq 0$ then $v(\hat{p}^0)+u(\bar{x}^0) > 0$ which is a contradiction. If $\langle \hat{p}^0, \bar{x}^0\rangle > 0$ this would contradict the feasibility of $\hat{p}^0$ or $\bar{x}^0$ by Theorem 1 of [6].

Let us now relax the conditions on $u$ from being closed strictly quasiconcave to being closed quasiconcave. This generalization can cause problems when determining the primal solution corresponding to the optimal dual solution. The difficulty occurs in using the utility transform to obtain the vector $x^0$ from $p^0$. There are two cases:

(a) $\{0\} \cap \partial^{\mathrm{loc}} u(x^0) = \emptyset$. In this case the LAGRANGE multiplier associated with the utility transform $\lambda > 0$ and $x^0$ can be determined by the relation

$$\lambda p^0 \in \partial^{\mathrm{loc}} u(x^0).$$

(b) $\{0\} \subset \partial^{\mathrm{loc}} u(x^0)$. In ths case $\lambda$ could be zero. To determine $x^0$ we must find $x$ vectors which satisfy

$$u(x^0) \geqq -v(p^0), \quad \langle x^0, p^0\rangle \leqq 0, \quad x^0 \in U$$
$$f_i(x^i) \geqq 0, \quad x^i \in F_i, \quad i=1, 2, \ldots, m$$
$$x \in X.$$

For those constraints which are binding $f_i(x^i) \geqq 0$, $x^i \in F_i$ can be replaced by

$$x^i \in \partial h(p^i).$$

This set is a convex set. While this is a weaker result than might be hoped for it should be noted that duality has given the value of the objective function $(-v(p^0))$ and the dual variables. From the work of ARROW and ENTHOVEN [5] this result is to be expected.

### 4. An Economic Example

Let us consider $u(x^0)$ being an aggregate utility function for a society. The utility function can reasonably be assumed to be quasiconcave and that if $\bar{x}^0 > x^0$ then $u(\bar{x}^0) > u(x^0)$. These are sufficient conditions for us to be able to apply the utility transform. A common utility function is

$$u(x^0) = \sum_i \alpha_i \log (x_i^0 + e_i)$$

where $\alpha_i$ and $e_i$ are parameters, $\alpha_i$ is non-negative and $e_i$ is the endowment of the $i$-th good.

Within the society there are various firms with production functions of the form

$$f_i(x^i) \geqq 0, \quad i = 1, 2, \ldots, m .$$

These functions can be reasonably assumed to be quasiconcave and positively homogeneous of degree zero. $x^i$, $i = 0, 1, 2, \ldots$ are the inputs and outputs of the society ($i = 0$) and its firms ($i = 1, \ldots, m$).

Each segment $x^i$-represents a different good or service with the sign being positive for input and negative for output. There are various implicit constraints on the values that the $x^i$ can take. These restrictions are captured in the set $U$ for the society and in $F_i$ for the $i$ th firm.

The conservation of goods and services require that the decision vectors $x^i$ be related by a simple subspace

$$X = \left\{ x \mid \sum_{i=0}^{m} x^i = 0 \right\}$$

A common production function used is the COBB-DOUGLAS. Let $x_1^i$ be labour and $x_2^i$ be capital. Then the production function for the $j$ th good produced by the $i$ th technology is

$$c_i \left( \frac{-x_1^i}{x_{j_i}^i} \right)^{a_i} \left( \frac{-x_2^i}{x_{j_i}^i} \right)^{1-a_i} \geqq 1$$

with

$$x_1^i \geqq 0, \quad x_2^i \geqq 0, \quad x_{j_i}^i \leqq 0 \quad \text{and}$$
$$x_k^i = 0, \quad k \neq 1, 2, \text{ or } j_i$$

where

$$c_i > 0 \quad \text{and} \quad 0 \leqq a_i \leqq 1 .$$

Combining the past few paragraphs we obtain a mathematical program with the form of program A.

Program C: Maximize $\sum_i \alpha_i \log (x_i^0 + e_i)$

subject to

$$c_i \left( \frac{-x_1^i}{x_{j_i}^i} \right)^{a_i} \left( \frac{-x_2^i}{x_{j_i}^i} \right)^{1-a_i} \geqq 1, \quad i = 1, 2, \ldots, m$$

$$x^0 \in \{x^0 \mid x^0 \geqq e\}$$
$$x^i \in \{x^i \mid x_1^i \geqq 0, \quad x_2^i \geqq 0, \quad x_{j_i}^i \leqq 0 \quad \text{and} \quad x_k^i = 0, \quad k \neq 1, 2, \text{ or } j_i\}, \qquad i = 1, 2, \ldots, m$$

$$\sum_{i=0}^{m} x^i = 0 .$$

To determine the dual we must first calculate a utility transform, a profit transform and a dual cone. The utility transform of $\sum_i \alpha_i \log (x_i^0 + e_i)$ is:

$$v(p^0) = \inf_{x^0 \geqq e} \{-\sum_i \alpha_i \log (x_i^0 + e_i) \mid \langle p^0, x^0 \rangle \leqq 0\}.$$

The infimum is attained when

$$\lambda p_i^0 = \frac{\alpha_i}{x_i^0 + e_i}, \quad \forall i$$

$$\langle p^0, x^0 \rangle = 0.$$

Thus

$$x_i^0 = -e_i + \frac{\alpha_i}{\lambda p_i^0}, \quad \forall i.$$

Since $\langle p^0, x^0 \rangle = 0$ we have

$$-\sum_i p_i^0 e_i + \frac{\sum_i \alpha_i}{\lambda} = 0.$$

Therefore

$$\lambda = \frac{\sum_i \alpha_i}{\sum_i p_i^0 e_i}$$

and

$$x_i^0 = -e_i + \frac{\alpha_i}{\sum_j \alpha_j} \frac{\sum_j p_j^0 e_j}{p_i^0}, \quad \forall i.$$

Substituting into the utility transform we obtain

$$v(p^0) = \sum_i \alpha_i \log \left( \frac{\sum_k \alpha_k}{\alpha_i} \cdot \frac{p_i^0}{\sum_k p_k^0 e_k} \right)$$

and

$$V = \{p^0 \mid p^0 \geqq 0\}.$$

The profit transform of

$$c_i \left( \frac{-x_1^i}{x_{j_i}^i} \right)^{\alpha_i} \left( \frac{-x_2^i}{x_{j_i}^i} \right)^{1-\alpha_i} \geqq 1$$

is more difficult. We appeal to the duality theory for posynomial programs of DUFFIN, PETERSON and ZENER [11].

$$h(p^i) = \inf \left\{ -\langle p^i, x^i \rangle \mid c_i \left( \frac{-x_1^i}{x_{j_i}^i} \right)^{\alpha_i} \left( \frac{-x_2^i}{x_{j_i}^i} \right)^{1-\alpha_i} \geqq 1 \right\}.$$

Since the production function is positively homogeneous of degree zero the infimum is either zero or $-\infty$. We are concerned with the former situation. Let

$$t_1 = \frac{-x_1^i}{x_{j_i}^i} \geqq 0 \quad \text{and} \quad t_2 = \frac{-x_2^i}{x_{j_i}^i}.$$

The infimum is equivalent to finding the

$$\inf \left(\frac{p_1^i}{p_{j_i}^i}\right) t_1 + \left(\frac{p_2^i}{p_{j_i}^i}\right) t_2 - 1$$

subject to

$$c_i (t_1)^{a_i} (t_2)^{1-a_i} \geqq 1 \; .$$

This formulation is obtained by dividing the objective function by $p_{j_i}^i x_{j_i}^i$. Neglecting the constant $-1$ in the objective function the problem may be written as a posynomial program with degree of difficulty zero.

$$\inf \left(\frac{p_1^i}{p_{j_i}^i}\right) t_1 + \left(\frac{p_2^i}{p_{j_i}^i}\right) t_2$$

subject to

$$(c_i)^{-1} (t_1)^{-a_i} (t_2)^{a_i - 1} \leqq 1 \; .$$

The geometric programming dual is

$$\sup \left(\frac{p_1^i}{p_{j_i}^i \delta_1}\right)^{\delta_1} \left(\frac{p_2^i}{p_{j_i}^i \delta_2}\right)^{\delta_2} \left(\frac{1}{c_i}\right)^{\delta_3}$$

subject to

$$\begin{aligned}
&\delta_1 + \delta_2 = 1 \\
&\delta_1 - a_i \delta_3 = 0 \\
&\delta_2 - (1 - a_i)\, \delta_3 = 0 \\
&\delta_1, \delta_2, \delta_3 \geqq 0 \, .
\end{aligned}$$

The constraints limit the dual feasible region to the point

$$\begin{aligned}
&\delta_1 = a_i \\
&\delta_2 = 1 - a_i \\
&\delta_3 = 1 \, .
\end{aligned}$$

Thus the value of the primal and dual objective functions at the optimum is

$$\left(\frac{p_1^i}{a_i p_{j_i}^i}\right)^{a_i} \left(\frac{p_2^i}{(1-a_i)\, p_{j_i}^i}\right)^{1-a_i} (c_i)^{-1}$$

For the infimum to be finite for the profit transform

$$\left(\frac{p_1^i}{a_i p_{j_i}^i}\right)^{a_i} \left(\frac{p_2^i}{(1-a_i)\, p_{j_i}^i}\right)^{1-a_i} (c_i)^{-1} - 1 \geqq 0 \, .$$

Thus the profit transform is

$$h(p^i) = 0$$

and

$$H = \left\{ p^i \mid \left(\frac{p_1^i}{a_i p_{j_i}^i}\right)^{a_i} \left(\frac{p_2^i}{(1-a_i)\, p_{j_i}^i}\right)^{1-a_i} \geqq c_i \right\}.$$

The dual cone of $X$ is

$$\left\{p \mid \sum_{i=0}^{m} \langle p^i, x^i \rangle \leqq 0, \ \forall x \in X\right\} = \{p \mid p^0 = p^i, \ \forall i\} .$$

Combining these results together the dual to program C is

Program D:

$$\text{Maximize } \sum_i \alpha_i \log \left( \frac{\sum_k \alpha_k}{\alpha_i} \cdot \frac{p_i^0}{\sum_k p_k^0 e_k} \right)$$

subject to

$$p^0 \geqq 0$$

$$\left(\frac{p_1^i}{a_i p_{j_i}^i}\right)^{a_i} \left(\frac{p_2^i}{(1-a_i)\, p_{j_i}^i}\right)^{1-a_i} \geqq c_i, \quad i = 1, \ldots, m$$

$$p^0 = p^i, \quad \forall i .$$

Program D can be reduced to

Program E:

$$\text{Maximize } \sum_i \alpha_i \log \left( \frac{\sum_k \alpha_k}{\alpha_i} \cdot \frac{p_i^0}{\sum_k p_k^0 e_k} \right)$$

subject to

$$\left(\frac{p_1^0}{a_i p_{j_i}^0}\right)^{a_i} \left(\frac{p_2^0}{(1-a_i)\, p_{j_i}^0}\right)^{1-a_i} \geqq c_i, \quad i = 1, \ldots, m$$

$$p^0 \geqq 0 .$$

**Acknowledgement**

The authors wish to acknowledge the helpful suggestions of the referees.

**References**

[1] MANGASARIAN, O. L.: Nonlinear Programming. McGraw-Hill. 1969.
[2] FENCHEL, W.: Convex Cones, Sets and Functions. Lecture Notes. Princeton University, 1953.
[3] ROCKAFELLAR, R. T.: Convex Analysis. Princeton University Press. 1970.
[4] PETERSON, E. L.: Geometric Programming. SIAM Review **18** (1976), 1–52.
[5] ARROW, K. J., and A. C. ENTHOVEN: Quasi-Concave Programming. Econometrica **29** (1967), 779–800.
[6] JEFFERSON, T. R., and C. H. SCOTT: A Duality Theory for Quasi-Concave Programs. Applicable Analysis **10** (1980), 165–177.
[7] GREENBERG, H. J., and W. P. PIERSKALLA: Quasi-Conjugate Function and Surrogate Duality. Cahiers du Centre d'Etudes de Recherche Opérationelle **15** (1973), 437–448.

[8] Crouzeix, J. P.: Conjugacy in Quasiconvex Analysis. Lecture Notes in Economics and Mathematical Systems **144**, Springer-Verlag, 1977.
[9] Jorgenson, D. W., and L. J. Lau: The Duality of Technology and Economic Behaviour. Review of Economic Studies **41** (1974), 181–200.
[10] Jacobsen, S. E.: Production Correspondence. Econometrica **38** (1970) 754–771.
[11] Duffin, R. J., E. L. Peterson and C. Zener: Geometric Programming. John Wiley and Sons, Inc., 1967.

**Zusammenfassung**

Dualitätsbetrachtungen sind ein wesentlicher Teil der Theorie der konkaven Optimierung. Grundlage der Dualität ist dabei die konkavkonjugierte Transformation. Dies wurde beispielsweise bei der Entwicklung der verallgemeinerten geometrischen Optimierung deutlich. Viele der gegenwärtigen Untersuchungen zur Dualitätstheorie zielen auf eine Abschwächung der Konkavitätsforderung ab. Hier wird eine Dualitätstheorie für Optimierungsprobleme mit quasikonkaver Zielfunktion und explizit-quasikonkaven Nebenbedingungen entwickelt. Es wird eine Verallgemeinerung der konkaven konjugierten Transformation eingeführt, die anstelle konkaver Funktionen quasikonkave Funktionen einander zuordnet. Es werden Optimalitätsbedingungen hergeleitet, die sich auf die Beziehungen des primalen quasikonkaven Optimierungsproblems zu seinem dualen Problem gründen. Diese Dualitätstheorie ist durch gewisse Probleme aus der mathematischen Ökonomie begründet und zielt auch auf diese ab. Eine Anwendung in der Ökonomie wird beschrieben.

**Резюме**

Рассмотрение соотношений двойственности является важным элементом теории вогнутого программирования. Основой двойственности при этом является вогнутое сопряженное преобразование. Например стало видно в развитие обобщенного геометрического программирования. Многие из настоящих исследований в теории двойственности преследуют цель ослабить требование вогнутости. Здесь развиваем теорию двойственности для задач программирования с квазивогнутой целевой функцией и явно-квазивогнутыми ограничениями. Вводится обобщение вогнутого сопряженного преобразования которое вместо вогнутых функции ставит в соответствие квазивогнутые функции. Производятся условия оптимальности основанные на соотношения между заданной квазивогнутой задачей программирования и двойственной ей задачей. Эта теория двойственности основана на некоторые задачи математической экономии и прицеливается на них. Описывается одно применение в экономии.

Received August 1979; revised February 1980

Math. Operationsforsch. Statist., Ser. Optimization, Vol. 11 (1980) No. 4, 531–536

# Eine hinreichende Bedingung für die Existenz eines eindeutigen lokalen Extremums[1]

JAN PALATA[2]

**Zusammenfassung:** In der Arbeit werden hinreichende Bedingungen für die Existenz einer lokalen Lösung eines nichtlinearen (im allgemeinen nichtkonvexen) Optimierungsproblems für einmal stetig differenzierbare Funktionen angegeben.

Der Gegenstand unserer Untersuchung ist die Frage der Existenz eines lokalen Extremums in der nichtlinearen (nichtkonvexen) Optimierung:

$$\min_{x \in N} F(x) ,$$

wobei (1)

$$N = \{x \in E_n \mid G_i(x) \leqq 0,\ i = 1, \ldots, m\} .$$

Wir stellen an dieses Problem folgende Forderungen: Die Funktionen $F(x)$, $G_i(x)$ ($i=1, \ldots, m$) haben stetige partielle Ableitungen erster Ordnung, und es gilt $\nabla F(x) \neq 0$, $\nabla G_i(x) \neq 0$ ($i=1, \ldots, m$).

Für die lokale Approximation der Menge $A$ im Punkt $x_0 \in \bar{A}$[3] werden wir den sog. Berührungskegel $S(x_0, A)$, der in [1] eingeführt wurde, benutzen (siehe Definition 5).

Es sei $p(x_0; v)$ die Halbgerade, die durch den Punkt $x_0 \in E_n$ und den Vektor $v \neq 0$ definiert ist, d. h. $p(x_0; v) = \{x \in E_n \mid x = x_0 + tv,\ t \in (0, \infty)\}$.

**Definition 1:** Falls $P(x_0; v)$ ein beliebiger polyedrischer Kegel in $E_n$ mit den Eigenschaften

a) $\dim P(x_0; v) = n$ ,
b) $x_0$ ist der einzige Scheitel von $P(x_0; v)$,
c) $p(x_0; v) \subset \operatorname{int} P(x_0; v)$

ist, so heißt die Menge

$$U(x_0; v) = \operatorname{int} P(x_0; v)$$

eine *polyedrische Umgebung* der Halbgeraden $p(x_0; v)$ in $E_n$.

---

[1] Resultate wurden auf der Tagung „Mathematische Optimierung-Theorie und Anwendungen, Eisenach 1978, vorgetragen.

[2] Matematicko-fyzikální fakulta University Karlovy, 11800 Praha 1, Malostranské nam. 25, ČSSR.

[3] $\bar{A}$ bedeutet die Abschließung von $A$.

**Definition 2:** Es sei $U(\boldsymbol{x}_0; \boldsymbol{v})$ eine beliebige polyedrische Umgebung der Halbgeraden $p(\boldsymbol{x}_0; \boldsymbol{v})$ und $H$ ein offener Halbraum in $E_n$ mit dem Rand $R_H$ und mit den Eigenschaften
1. $\boldsymbol{x}_0 \in H$ ,
2. $U(\boldsymbol{x}_0; \boldsymbol{v}) \cap R_H \neq \emptyset$ und beschränkt,
dann heißt die Menge

$$\gamma(H; P) = U(\boldsymbol{x}_0, \boldsymbol{v}) \cap R_H$$

ein *echter Schnitt* der polyedrischen Umgebung $U(\boldsymbol{x}_0; \boldsymbol{v})$ von $p(\boldsymbol{x}_0; \boldsymbol{v})$.

**Definition 3:** Ein echter Schnitt $\gamma'(H'; P)$ von $U(\boldsymbol{x}_0; \boldsymbol{v})$ heißt *feiner* als ein anderer echter Schnitt $\gamma(H; P)$ von derselben Umgebung $U(\boldsymbol{x}_0; \boldsymbol{v})$, falls

$$\gamma'(H'; P) \subset (H \cap U(\boldsymbol{x}_0; \boldsymbol{v}))$$

gilt.

**Bemerkung 1:** Falls $\gamma'(H'; P)$ feiner als $\gamma(H; P)$ ist, so schreiben wir dafür

$$\gamma'(H'; P) \lessdot \gamma(H; P) .$$

**Definition 4:** Es sei $\boldsymbol{x}_0$ ein festgewählter Punkt einer gegebenen Menge $\bar{A} \subset E_n$. Eine Halbgerade $p(\boldsymbol{x}_0; \boldsymbol{v})$ mit der Eigenschaft, daß für jede polyedrische Umgebung $U(\boldsymbol{x}_0; \boldsymbol{v})$ von $p(\boldsymbol{x}_0; \boldsymbol{v})$ ein echter Schnitt $\gamma(H; P)$ in der Weise existiert, daß $\gamma'(H'; P) \cap A \neq \emptyset$ für alle $\gamma'(H'; P) \lessdot \gamma(H; P)$ gilt, heißt eine *σ-Halbgerade* der Menge $A$ im Punkt $\boldsymbol{x}_0 \in \bar{A}$.

**Definition 5:** Es sei $\boldsymbol{x}_0$ ein festgewählter Punkt einer gegebenen Menge $\bar{A} \subset E_n$ und $\sum_\sigma p(\boldsymbol{x}_0; \boldsymbol{v})$ die Menge aller σ-Halbgeraden der Menge $A$ im Punkt $\boldsymbol{x}_0$. Die Menge

$$S(\boldsymbol{x}_0; A) = \{x_0\} \cup \sum_\sigma p(\boldsymbol{x}_0; \boldsymbol{v})$$

heißt der *Berührungskegel* von $A$ im Punkt $\boldsymbol{x}_0$.

Wir definieren die Mengen

$$M = \{\boldsymbol{x} \in E_n \mid F(\boldsymbol{x}) \leqq 0\}, \quad N_i = \{\boldsymbol{x} \in E_n \mid G_i(\boldsymbol{x}) \leqq 0\} \quad (i = 1, \ldots, m),$$
$$I = \{i \in \{1, \ldots, m\} \mid G_i(\boldsymbol{x}_0) = 0\} .$$

$\partial A$ bezeichne den Rand der Menge $A$. O. B. d. A. können wir voraussetzen, daß im Punkt eines lokalen Extremums $\boldsymbol{x}_0$ die Beziehung $F(\boldsymbol{x}) = 0$ gilt, d. h. $\boldsymbol{x}_0 \in \partial M$. Der Punkt $\boldsymbol{x}_0 \in M \cap N$ ist offenbar ein lokales Minimum (eindeutiges lokales Minimum) der Funktion $F(\boldsymbol{x})$ in bezug auf die Menge $N$ genau dann, wenn die Mengen $M$ (int $M$) und $N$ im Punkt $\boldsymbol{x}_0$ lokal disjunkt sind, d. h. wenn eine Umgebung $U(\boldsymbol{x}_0)$ mit der Eigenschaft $U(\boldsymbol{x}_0) \cap \text{int } M \cap N = \emptyset$ $(U(\boldsymbol{x}_0) \cap M \cap N = \{\boldsymbol{x}_0\})$ existiert. Die Bedingung $S(\boldsymbol{x}_0; N) \cap \text{int } S(\boldsymbol{x}_0; M) = \emptyset$ ist ein notwendiges Kriterium für die Existenz eines lokalen Minimums der Aufgabe (1). Dieses Resultat enthält der folgende Satz.

**Satz 1** (siehe [2]): *Ist $M \cap N \neq \emptyset$, dann gilt*

$$\boldsymbol{x}_0 \in M \cap N, \quad S(\boldsymbol{x}_0; N) \cap \text{int } S(\boldsymbol{x}_0; M) \neq \emptyset \Rightarrow$$
$$U(\boldsymbol{x}_0) \cap N \cap \text{int } M \neq \emptyset \quad \textit{für jede Umgebung } U(\boldsymbol{x}_0) .$$

Bevor wir eine hinreichende Bedingung für ein eindeutiges lokales Minimum formulieren, geben wir zwei Hilfssätze an.

**Lemma 1** (siehe [2]): *Es seien* $\Sigma$ *ein polyedrischer Kegel mit dem Scheitel* $\boldsymbol{x}_0$ *und* $N=\left\{\boldsymbol{x}\in E_n \mid \sum_{\alpha=1}^{n} a^\alpha (x^\alpha - x_0^\alpha)=0\right\}$ *eine Hyperebene in* $E_n$ *mit den Eigenschaften* $\Sigma\cap N=\{\boldsymbol{x}_0\}$, $\Sigma\setminus\{\boldsymbol{x}_0\}\subset H^+$, *wobei* $H^+=\left\{\boldsymbol{x}\in E_n \mid \sum_{\alpha=1}^{n} a^\alpha (x^\alpha - x_0^\alpha)>0\right\}$. *Dann existiert ein Rotationskegel* $K_{\boldsymbol{a}}(\boldsymbol{x}_0;\varepsilon)$ *mit dem Scheitel* $\boldsymbol{x}_0$ *und mit der Achse, die durch den Vektor* $\boldsymbol{a}$ *definiert ist, so daß*

1) $\Sigma\subset K_{\boldsymbol{a}}(\boldsymbol{x}_0;\varepsilon)$

2) $K_{\boldsymbol{a}}(\boldsymbol{x}_0;\varepsilon)\setminus\{\boldsymbol{x}_0\}\subset H^+$

gilt.

**Bemerkung 2:** Der Kegel $K_{\boldsymbol{a}}(\boldsymbol{x}_0;\varepsilon)$ im Lemma 1 wird durch die Ungleichung

$$\|\boldsymbol{x}-\boldsymbol{x}_0\| - \sqrt{1+\frac{1}{\varepsilon^2}}\sum_{\alpha=1}^{n} a^\alpha (x^\alpha - x_0^\alpha) \leqq 0 \text{ charakterisiert.}$$

**Lemma 2** (siehe [2]): *Es sei* $\partial M=\{\boldsymbol{x}\in E_n \mid F(\boldsymbol{x})=0\}$ *eine glatte Hyperfläche in* $E_n$ *und* $\boldsymbol{x}_0\in\partial M$ *beliebig. Dann sind für jedes* $\varepsilon>0$ *die Mengen* $\partial M$ *und* $K_{\nabla F(\boldsymbol{x}_0)}(\boldsymbol{x}_0;\varepsilon)$ *im Punkt* $\boldsymbol{x}_0$ *lokal disjunkt.*

**Satz 2:** *Wenn* $\bigcap_{i\in I} S(\boldsymbol{x}_0;N_i)\cap S(\boldsymbol{x}_0;M)=\{x_0\}$ *erfüllt ist, gibt es eine Umgebung* $U(\boldsymbol{x}_0)$ *mit* $U(\boldsymbol{x}_0)\cap M\cap N=\{\boldsymbol{x}_0\}$.

Beweis: Nichttrivial ist der Fall $\bigcap_{i\in I} S(\boldsymbol{x}_0;N_i)\neq\{\boldsymbol{x}_0\}$. Es sei dann $K_{\nabla F(\boldsymbol{x}_0)}(\boldsymbol{x}_0;\varepsilon)$ ein Rotationskegel mit $\left(\bigcap_{i\in I} S(\boldsymbol{x}_0;N_i)\setminus\{\boldsymbol{x}_0\}\right)\subset\operatorname{int} K_{\nabla F(\boldsymbol{x}_0)}(\boldsymbol{x}_0;\varepsilon)$ und $r\in I$ beliebig, aber fest gewählt. Für $\boldsymbol{x}_0+\boldsymbol{h}\in T(\boldsymbol{x}_0)$, $T(\boldsymbol{x}_0)=\left\{\boldsymbol{x}\in E_n \mid \sum_{\alpha=1}^{n}\left(\frac{\partial F}{\partial x^\alpha}\right)_{\boldsymbol{x}_0}(x^\alpha - x_0^\alpha)=0\right\}$ bezeichnen wir mit $p_{\boldsymbol{h}}$ die Gerade $\{\boldsymbol{x} \mid \boldsymbol{x}=\boldsymbol{x}_0+\boldsymbol{h}+t\nabla F(\boldsymbol{x}_0),\ t\in(-\infty,\infty)\}$.

a) Die Gerade $p_{\boldsymbol{0}}$ liegt nicht in der tangentialen Hyperebene $T_r(\boldsymbol{x}_0)$ der Hyperfläche $\partial N_r$ im Punkt $\boldsymbol{x}_0$.

Wählen wir orthogonale Vektoren $\boldsymbol{a}_1,\ldots,\boldsymbol{a}_{n-1}$ mit $\boldsymbol{x}_0+\boldsymbol{a}_i\in T(\boldsymbol{x}_0)$, $i=1,\ldots,n-1$, dann gibt es eine Funktion $g_r(\xi^1,\ldots,\xi^{n-1})$, so daß die Gleichung $G_r\left(\boldsymbol{x}_0+\sum_{i=1}^{n-1}\xi^i\boldsymbol{a}_i+\xi^n\nabla F(\boldsymbol{x}_0)\right)=0$ in einer gewissen Umgebung $\tilde{U}_r(\boldsymbol{x}_0)$ mit $\xi^n=g_r(\xi^1,\ldots,\xi^{n-1})$ äquivalent ist. In dem Koordinatensystem, das durch den Punkt $\boldsymbol{x}_0$ und die Vektoren $\boldsymbol{a}_1,\ldots,\boldsymbol{a}_{n-1},\nabla F(\boldsymbol{x}_0)$ gegeben ist, hat die Gleichung der tangentialen Hyperebene $T_r(\boldsymbol{x}_0)$ die Gestalt $\xi^n=\sum_{i=1}^{n-1}\left(\frac{\partial g_r}{\partial\xi^i}\right)_0\xi^i$, $\xi^i\in(-\infty,\infty)$, $(i=1,\ldots,n-1)$ und die Gleichung der Geraden $p_{\boldsymbol{h}}$ die Form $\xi^i=\bar{h}^i$ $(i=1,\ldots,n-1)$, $\xi^n$ beliebig, wobei $(\bar{h}^1,\ldots,\bar{h}^{n-1},0)$ die Koordinaten des Punktes $\boldsymbol{x}_0+\boldsymbol{h}$ in diesem neuen Koordinatensystem sind. Wir wählen $\boldsymbol{h} \neq 0$ derart, daß $\boldsymbol{x}_0 + \boldsymbol{h} \in \tilde{U}_r(\boldsymbol{x}_0)$ erfüllt ist und daß die Gerade $p_h$ den Teil des Kegels $\bigcap_{i \in I} S(\boldsymbol{x}_0, N_i)$, der in der Hyperebene $T_r(\boldsymbol{x}_0)$ liegt, schneidet. Es ist

$$\frac{\sum_{i=1}^{n-1} \left(\frac{\partial g_r}{\partial \xi^i}\right)_0 \bar{h}^i}{\|\boldsymbol{h}\|} > \operatorname{tg}\left(\frac{\pi}{2} - \varphi\right)$$

($2\varphi$ ist der Scheitelwinkel des Kegels $K_{\nabla F(\boldsymbol{x}_0)}(\boldsymbol{x}_0; \varepsilon)$) und deshalb auch $\frac{g_r(\bar{\boldsymbol{h}})}{\|\bar{\boldsymbol{h}}\|} > \operatorname{tg}\left(\frac{\pi}{2} - \varphi\right)$ für kleine $\|\bar{\boldsymbol{h}}\|$. Das bedeutet, daß in einer gewissen Umgebung $U_r(\boldsymbol{x}_0) \setminus \{\boldsymbol{x}_0\}$ der Teil der Menge $\partial N$, der zur Hyperfläche $\partial N_r$ gehört, in der Menge $\operatorname{int} K_{\nabla F(\boldsymbol{x}_0)}(\boldsymbol{x}_0; \varepsilon)$ liegt.

b) Wenn die Gerade $p_0$ in der Hyperebene $T_r(\boldsymbol{x}_0)$ liegt, wählen wir den Vektor $\boldsymbol{a}$ so, daß $\sum_{\alpha=1}^{n} \left(\frac{\partial F}{\partial x^\alpha}\right)_{\boldsymbol{x}_0} a^\alpha > 0$ gilt und die Gerade $l_0 := \{\boldsymbol{x} \in E_n \mid \boldsymbol{x} = \boldsymbol{x}_0 + t\boldsymbol{a},\ t \in (-\infty, \infty)\}$ nicht in der Hyperebene $T_r(\boldsymbol{x}_0)$ liegt. Sind $\boldsymbol{a}_1, \ldots, \boldsymbol{a}_{n-1}$ orthonormale Vektoren mit $\boldsymbol{x}_0 + \boldsymbol{a}_i \in T(\boldsymbol{x}_0)$ $(i = 1, \ldots, n-1)$, dann kann man in dem Koordinatensystem $(\boldsymbol{x}_0; \boldsymbol{a}_1, \ldots, \boldsymbol{a}_{n-1}, \boldsymbol{a})$ lokal die Hyperfläche $\partial N_r$ folgendermaßen beschreiben:

$$\mu^n = g_r(\mu^1, \ldots, \mu^{n-1})\,.$$

Die Gerade $l_h = \{\boldsymbol{x} \in E_n \mid \boldsymbol{x} = \boldsymbol{x}_0 + \boldsymbol{h} + t\boldsymbol{a},\ t \in (-\infty, \infty)\}$ hat dann folgende Beschreibung: $\mu^1 = \tilde{h}^1, \ldots, \mu^{n-1} = \tilde{h}^{n-1}$, $\mu^n$ beliebig, wobei $(\tilde{h}^1, \ldots, \tilde{h}^{n-1}, 0)$ die neuen Koordinaten des Punktes $\boldsymbol{x}_0 + \boldsymbol{h}$ sind. Wegen der Lage der Geraden $p_0$ gibt es die Zahlen $c^1, \ldots, c^{n-1}$, so daß die Gleichung der tangentialen Hyperebene $T_r(\boldsymbol{x}_0)$ die Form

$$\mu^n = -\frac{\sum_{i=1}^{n-1} c^i \mu^i}{\sum_{i=1}^{n-1} c^i \boldsymbol{a}_i \boldsymbol{a}}$$

hat. Wir wählen $\boldsymbol{x}_0 + \boldsymbol{h} \in T(\boldsymbol{x}_0) \setminus \{\boldsymbol{x}_0\}$ so, daß die Gerade $l_h$ die Menge $\bigcap_{i \in I} S(\boldsymbol{x}_0; N_i) \cap T_r(\boldsymbol{x}_0)$ schneidet. Wenn der Schnittpunkt der Geraden $l_h$ mit dem Kegel $K_{\nabla F(\boldsymbol{x}_0)}(\boldsymbol{x}_0; \varepsilon)$, für den $(\boldsymbol{x}_0 + \boldsymbol{h}, \boldsymbol{x}_0 + \boldsymbol{h} + t\boldsymbol{a}) \subset \operatorname{ext} K_{\nabla F(\boldsymbol{x}_0)}(\boldsymbol{x}_0; \varepsilon)$ gilt, dem Wert $t = t_1$ des Parameters der Geraden $l_h$ entspricht, so gilt $g_r(\tilde{h}) - t_1 > 0$ für genügend kleine $\|\tilde{\boldsymbol{h}}\|$.

Also ist bewiesen, daß es eine Umgebung $O(\boldsymbol{x}_0)$ mit $(O(\boldsymbol{x}_0) \cap \partial N \setminus \{\boldsymbol{x}_0\}) \supset \operatorname{int} K_{\nabla F(\boldsymbol{x}_0)}(\boldsymbol{x}_0; \varepsilon)$ gibt. Um $(N \cap O(\boldsymbol{x}_0) \setminus \{\boldsymbol{x}_0\}) \subset \operatorname{int} K_{\nabla F(\boldsymbol{x}_0)}(\boldsymbol{x}_0; \varepsilon)$ zu zeigen, genügt es zu beweisen, daß alle $\boldsymbol{z} \in O(\boldsymbol{x}_0) \cap N$ mit der Eigenschaft $G_i(\boldsymbol{z}) < 0$ für jedes $i \in I$ in der Menge $\operatorname{int} K_{\nabla F(\boldsymbol{x}_0)}(\boldsymbol{x}_0; \varepsilon)$ liegen. Es sei also $\boldsymbol{z}$ ein solcher Punkt mit $G_i(\boldsymbol{z}) < 0$ $(i \in I)$, und es sei $\boldsymbol{z} \notin \operatorname{int} K_{\nabla F(\boldsymbol{x}_0)}(\boldsymbol{x}_0; \varepsilon)$. Unter dieser Voraussetzung muß ein Index $i_0 \in I$ mit $\sum_{\alpha=1}^{n} \left(\frac{\partial G_{i_0}}{\partial x^\alpha}\right)_{\boldsymbol{x}_0} (x^\alpha - x_0^\alpha) > 0$ existieren, woraus die Existenz der Zahl $\delta > 0$ folgt, so daß für $\lambda \in (0, \delta)$ $G_{i_0}(\boldsymbol{x}_0 + \lambda(\boldsymbol{z} - \boldsymbol{x}_0)) > 0$ gilt.

Da die Funktion $G_{i_0}(\boldsymbol{x})$ stetig ist, liegt auf dem Segment $\langle \boldsymbol{x}_0, \boldsymbol{z} \rangle$ ein Punkt $\boldsymbol{y}$, in dem die Funktion $G_{i_0}(\boldsymbol{x})$ verschwindet, und deshalb ist $\boldsymbol{y} \in \operatorname{int} K_{\nabla F(\boldsymbol{x}_0)}(\boldsymbol{x}_0; \varepsilon)$. Somit erhalten wir einen Widerspruch, denn jede offene Halbgerade mit dem Anfangspunkt $\boldsymbol{x}_0$, welcher der Scheitel des konvexen Kegels $K_{\nabla F(\boldsymbol{x}_0)}(\boldsymbol{x}_0; \varepsilon)$ ist, kann nur ganz im Inneren des Kegels oder ganz außerhalb des Kegels oder auf dem Rand dieses Kegels liegen. Das gilt also auch für das Segment $(\boldsymbol{x}_0, \boldsymbol{z}\rangle$. Daraus ergibt sich, daß die Mengen $M$ und $N$ im Punkt $\boldsymbol{x}_0$ lokal disjunkt sind.

**Beispiel:**

$$F(\boldsymbol{x}) = -x^2 + (x^1)^3, \quad G_1(\boldsymbol{x}) = x^2 - \operatorname{arctg} x^1, \quad G_2(\boldsymbol{x}) = x^2 + \frac{x^1}{3}$$

$$\boldsymbol{x}_0 = (0, 0)$$

$$S(\boldsymbol{x}_0; M) = \{\boldsymbol{x} \in E_2 \mid x^2 \geqq 0\}, \quad S(\boldsymbol{x}_0; N_1) = \{\boldsymbol{x} \in E_2 \mid x^2 - x^1 \leqq 0\}$$

$$S(\boldsymbol{x}_0; N_2) = \left\{\boldsymbol{x} \in E_2 \mid x^2 + \frac{x^1}{3} \leqq 0\right\}.$$

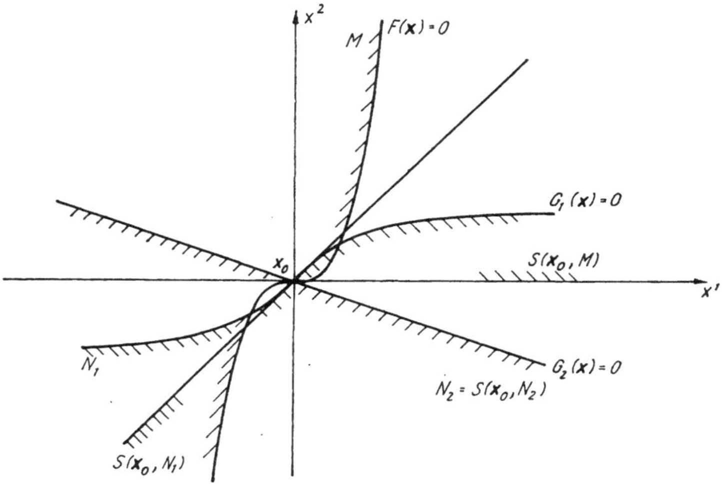


Abb. 1

Die Bedingung aus dem Satz 2 ist erfüllt. Der Punkt $\boldsymbol{x}_0 = (0, 0)$ ist ein eindeutiges lokales Minimum der Funktion $F(\boldsymbol{x})$ bezüglich der Menge $N$.

**Bemerkung 3:** Diese Theorie ist auch für Extremwertaufgaben ohne Restriktionen, anwendbar. Die Aufgabe ist dann mit dem Problem $\min \{z \in E_1 \mid z - F(\boldsymbol{x}) \leqq 0, -z + F(\boldsymbol{x}) \leqq 0\}$ äquivalent. Es gilt $\left(\frac{\partial z}{\partial x^1}, \ldots, \frac{\partial z}{\partial x^n}, \frac{\partial z}{\partial z}\right) = (0, \ldots, 0, 1)$, und es folgt aus dem Satz 1, daß $\nabla F(\boldsymbol{x}_0) = 0$. Das ist eine bekannte notwendige Bedingung. Die Bedingung aus dem Satz 2 wird aber in diesem Fall niemals erfüllt. Diese Tatsache bedeutet, daß die Approximationen erster Ordnung ohne weitere Voraussetzungen an unsere Funktionen zu grob sein können.

**Bemerkung 4:** Alle Untersuchungen betrafen lokale Eigenschaften. Es würde deshalb genügen, die obigen Voraussetzungen an die Funktionen $F(\boldsymbol{x}), G_i(\boldsymbol{x})$

($i=1, \ldots, m$) nur für eine Umgebung $U(\boldsymbol{x}_0)$ zu formulieren. Aus Gründen der einfachen Darstellung haben wir darauf verzichtet.

**Bemerkung 5:** Mittels dieser Technik formulierte BOLTJANSKIJ [3] analog notwendige Bedingungen. In der vorliegenden Arbeit wurde diese Methode auch für die Herleitung hinreichender Bedingungen benutzt. Der Satz 2 gilt gleichfalls mit anderen bekannten Approximationskegeln erster Ordnung, weil im betrachteten Fall die Kegel (oder eventuell ihre Abschließung oder Translation in den Punkt $\boldsymbol{x}_0$) meistens mit $S(\boldsymbol{x}_0; M) = \left\{\boldsymbol{x} \in E_n \mid \left(\frac{\partial F}{\partial x^\alpha}\right)_{\boldsymbol{x}_0} (x^\alpha - x_0^\alpha) \leqq 0\right\}$ identisch sind. Das ist z. B. der Fall für den tangentialen Kegel [4], den tangentialen Sukzessionskegel [5], das „Zelt" und das „lokale Zelt" der Menge [3], den Außenkegel, Simplexkegel [6], inneren Kegel [7], für die konvexe Approximation der Menge [7], den Ableitungskegel [8], die Kegelapproximation zweiter Art [4], den schwachen tangentialen Kegel [6], den Radialkegel (Kegel der zulässigen Richtungen) [4] usw.

### Literaturverzeichnis

[1] NOŽIČKA, F.: Über einfache Mannigfaltigkeiten in linearem affinen Raum $A_n^m$ in globaler Auffassung. Czech. Math. Journal **6** (1976).

[2] PALATA, J.: First-Order Necessary Condition for the Existence of Optimal Point in Nonlinear Programming Problem. Erscheint in Aplikace matematiky.

[3] BOLTJANSKIJ, V. G.: Metod šatrov v teorii ekstremalnych zadač, Uspechi matematičeskich nauk XXX, 000–000 (1975).

[4] CANON, M. D., CULLUM Jr., C. D., POLAK, E.: Theory of Optimal Control and Mathematical Programming. McGraw-Hill 1970.

[5] ABADIE, J.: Nonlinear Programming, North Holland Publishing Company 1967.

[6] WIERZBICKI, A. P.: Maximum Principle for Semiconvex Performance Functionals. SIAM J. Control. **10** (1972), 3.

[7] NEUSTADT, L. W.: An Abstract Variational Theory with Applications to a Broad Class of Optimization Problems. I. General Theory. J. SIAM Control. **4** (1966), 3.

[8] HESTENES, M. R.: Calculus of Variations and Optimal Control Theory. John Wiley 1966.

### Резюме

В этой работе даны достаточные условия для существования локального решения нелинейной (вообще не выпуклой) задачи программирования. Рассматриваются один раз непрерывно дифференцируемые функции.

### Summary

In the paper a sufficient condition for the existence of local unique solution of nonlinear (general non-convex) programming problem is given. Once continuously differentiable functions are considered.

Received January 1979; revised January 1980

Math. Operationsforsch. Statist., Ser. Optimization, Vol. 11 (1980) No. 4, 537–545

# Local Convergence of the Method of Multipliers[1]

JAROSLAV RAMÍK[2]

**Summary:** This paper deals with the generalized HESTENES-POWELL method of multipliers for solving constrained optimization problems. Augmented LAGRANGIANS which are not necessarily concave in the generalized penalty parameter are introduced. Only the case of equality constraints is considered and local convergence theorems are provided.

## 1. Introduction and Preliminaries

During last years, penalty function methods for solving nonlinear constrained optimization problems have gained recognition as one of the most effective class of methods. However, it is well known that these methods have an increasingly unfavorable structure due to ill-conditioning.

This difficulty is partly removed in methods, often called methods of multipliers, see e.g. [4], [7]. In these methods, the penalty term is added not to the objective function but to the LAGRANGIAN function which is then minimized. A sequence of unconstrained minimization problems is solved, but in contrast to pure penalty methods the penalty coefficients need not be increased to infinity, thus avoiding the associated extreme ill-conditioning. This paper deals with an analysis of some aspects of such methods mostly related to their local convergence.

In this paper we consider the following nonlinear constrained minimization problem:

$$\begin{aligned}&\text{minimize } f(\boldsymbol{x})\\&\text{subject to } \boldsymbol{g}(\boldsymbol{x})=0\ ,\end{aligned} \tag{1}$$

where $f: R^n \to R^1$, $\boldsymbol{g}: R^n \to R^m$, $m \leqq n$, $f$, $\boldsymbol{g}$ are twice continuously differentiable functions, generally nonconvex.

In our approach we consider a more general form of an augmented LAGRANGIAN. The *ordinary* LAGRANGIAN is defined by the usual manner:

$$F: R^n \times R^m \to R^1\ ,$$
$$F(x, \boldsymbol{\lambda}) = f(\boldsymbol{x}) + \boldsymbol{\lambda}^{\mathsf{T}} \boldsymbol{g}(\boldsymbol{x})\ .$$

---

[1] This paper was presented in a bit different form at the conference "Mathematische Optimierung – Theorie und Anwendungen", November 1978, Eisenach, DDR.

[2] Department of cybernetics and operational research, Charles University, 11800 Prague 1, Malostranské nám. 25, Czechoslovakia.

By a *penalty function* we mean the following function $P$:

$$P : R^m \times G \to R^1 ,$$

where $G$ is a metric space, e.g. $G$ is an $r$-dimensional EUCLIDean space $R^r$, $r \geqq 1$.

The *augmented* LAGRANGIAN $L$ is defined as a sum of the ordinary LAGRANGIAN $F$ and the penalty function $P$ as follows:

$$L : R^n \times R^m \times G \to R^1 ,$$
$$L(\boldsymbol{x}, \boldsymbol{\lambda}, \boldsymbol{A}) = F(\boldsymbol{x}, \boldsymbol{\lambda}) + P(\boldsymbol{g}(\boldsymbol{x}), \boldsymbol{A}) . \tag{2}$$

Any mapping $M$ such that

$$M : R^n \times R^m \times G \to G$$

is said to be *a penalty update formula*. We shall assume that the penalty function $P$ satisfies the following conditions:

(a) The function $P(\boldsymbol{y}, \boldsymbol{A})$ is twice continuously differentiable with respect to $\boldsymbol{y}$, moreover the mappings $\nabla_{\boldsymbol{y}} P$, $\nabla^2_{\boldsymbol{yy}} P$ are continuous on $R^m \times G$.

(b) $P(0, \boldsymbol{A}) = 0$ for all $\boldsymbol{A} \in G$.

(c) $\nabla_{\boldsymbol{y}} P(0, \boldsymbol{A}) = 0$ for all $\boldsymbol{A} \in G$.

By the symbols $\nabla_{\boldsymbol{y}} P$, $\nabla^2_{\boldsymbol{yy}} P$ we mean the gradient and the HESSIAN matrix of the function $P(\cdot, \boldsymbol{A})$, respectively.

Notice that the penalty function $P$ has a compact form. Thus, our case is not covered by the ARROW-GOULD-HOWE's paper [1] or MANGASARIAN's paper [5] where the augmented LAGRANGIANs have a separable form. There is also a remarkable fact that concavity in the penalty parameter $\boldsymbol{A}$ is not assumed (when $G$ is a linear space), so that ROCKAFELLAR's duality theory cannot be applied, see [9], [10]. We are, however, restricted only to the case of equality constraints. To treat inequality constraints one would have to assume additional properties of both the augmented LAGRANGIAN and the solution point. Nevertheless, inequality constraints may be treated either by supplying additional (slack) variables, see e.g. [3], or by eliminating nonbinding (inactive) constraints.

**Example 1.1:** $P(\boldsymbol{y}, \boldsymbol{A}) = \exp\left(\frac{1}{2} \boldsymbol{y}^\top \boldsymbol{A} \boldsymbol{y}\right) - 1$ where $\boldsymbol{y} \in R^m$, $\boldsymbol{A}$ is an $m \times m$-matrix.

In accordance with the known terminology we say that the $\boldsymbol{x}^* \in R^n$ is a *local solution of the problem* (1) if $\boldsymbol{g}(\boldsymbol{x}^*) = 0$ and there is a neighborhood $U$ of $\boldsymbol{x}^*$ such that $f(\boldsymbol{x}) \geqq f(\boldsymbol{x}^*)$ holds for all $\boldsymbol{x} \in U$ and $\boldsymbol{g}(\boldsymbol{x}) = 0$.

As it is obvious in a local convergence analysis we give some hypotheses concerning a solution point.

Assumption (A): A vector $\boldsymbol{x}^* \in R^n$ is an isolated local solution of the problem (1).

Assumption (B): Vectors $\boldsymbol{x}^* \in R^n$, $\boldsymbol{\lambda}^* \in R^m$ satisfy the following relations:

$$\boldsymbol{g}(\boldsymbol{x}^*) = 0, \quad \nabla_{\boldsymbol{x}} F(\boldsymbol{x}^*, \boldsymbol{\lambda}^*) = \nabla f(\boldsymbol{x}^*) + \nabla \boldsymbol{g}(\boldsymbol{x}^*)\, \boldsymbol{\lambda}^* = 0 . \tag{3}$$

The matrix $\nabla \boldsymbol{g}(\boldsymbol{x}^*)$ has full rank $m$.

$$\boldsymbol{h}^{\mathsf{T}} \nabla^2_{\boldsymbol{xx}} F(\boldsymbol{x}^*, \boldsymbol{\lambda}^*)\, \boldsymbol{h} > 0 \tag{4}$$

for all $0 \neq \boldsymbol{h} \in R^n$ such that $\nabla \boldsymbol{g}(\boldsymbol{x}^*)^{\mathsf{T}} \boldsymbol{h} = 0$.

The assumption (B) is the *second order sufficiency condition* for an isolated local solution of the problem (1). When the condition (B) is satisfied then (A) is also satisfied.

For any $c \in R^1$ we define sets $H(c)$, $H^0(c)$ as follows:

$$H(c) = \{\boldsymbol{A} \in G \mid \mu_1(\nabla^2_{\boldsymbol{yy}} P(0, \boldsymbol{A})) \geqq c\}\ ,$$
$$H^0(c) = \{\boldsymbol{A} \in G \mid \mu_1(\nabla^2_{\boldsymbol{yy}} P(0, \boldsymbol{A})) > c\}\ .$$

By the symbol $\mu_1(\boldsymbol{B})$ or $\mu_m(\boldsymbol{B})$ the smallest or the largest eigenvalue of the matrix $\boldsymbol{B}$ is understood.

**Example 1.2:** Let $P$ be as in the example 1.1., i.e.

$$P(\boldsymbol{y}, \boldsymbol{A}) = \exp\left(\frac{1}{2}\, \boldsymbol{y}^{\mathsf{T}} \boldsymbol{A} \boldsymbol{y}\right) - 1\ ,$$

where $\boldsymbol{A} \in G$ and $G = \{\boldsymbol{A} \mid \boldsymbol{A}$ is an $m \times m$-symmetric matrix$\}$. Then

$$\nabla^2_{\boldsymbol{yy}} P(0, \boldsymbol{A}) = \boldsymbol{A}\ ,$$
$$H(c) = \{\boldsymbol{A} \in G \mid \mu_1(\boldsymbol{A}) \geqq c\}\ .$$

## 2. Auxiliary Results

The following lemma is a modification of the known DEBREU's theorem, see e.g. [1]. We give also the proof as the proof from [1] cannot be applied.

**Lemma 2.1:** *Let $\boldsymbol{A}$ be an $n \times n$-matrix, $\boldsymbol{B}$ be an $n \times m$-matrix and let $\boldsymbol{x}^{\mathsf{T}} \boldsymbol{A} \boldsymbol{x} > 0$ for all $0 \neq \boldsymbol{x} \in R^n$ such that $\boldsymbol{B}\boldsymbol{x} = 0$.*

*Then there exist positive constants $\tau$, $c^*$ such that*

$$\boldsymbol{x}^{\mathsf{T}} (\boldsymbol{A} + \boldsymbol{B}^{\mathsf{T}} \boldsymbol{C} \boldsymbol{B})\, \boldsymbol{x} \geqq \tau \|\boldsymbol{x}\|^2 \tag{5}$$

*for all $\boldsymbol{x} \in R^n$ and all symetric $m \times m$-matrices $\boldsymbol{C}$ with $\mu_1(\boldsymbol{C}) \geqq c^*$.*

Proof: Clearly, it is sufficient to prove the lemma for all $\boldsymbol{x} \in S = \{\boldsymbol{x} \in R^m \mid \|\boldsymbol{x}\| = 1\}$. Suppose, conversely, that the conclusion of the lemma is false. Then for every integer $k$ there exist $\boldsymbol{x}_k \in S$ and a symmetric $m \times m$-matrix $\boldsymbol{C}_k$ with $\mu_1(\boldsymbol{C}_k) \geqq k$ such that

$$\boldsymbol{x}_k^{\mathsf{T}} (\boldsymbol{A} + \boldsymbol{B}^{\mathsf{T}} \boldsymbol{C}_k \boldsymbol{B})\, \boldsymbol{x}_k < \frac{1}{k}\ . \tag{6}$$

However, for all $\boldsymbol{y} \in R^m$ we have

$$\boldsymbol{y}^{\mathsf{T}} \boldsymbol{C}_k \boldsymbol{y} \geqq \mu_1(\boldsymbol{C}_k)\, \|\boldsymbol{y}\|^2\ ,$$

consequently,

$$\boldsymbol{x}^{\mathsf{T}}\boldsymbol{B}^{\mathsf{T}}\boldsymbol{C}_k\boldsymbol{B}\boldsymbol{x} \geqq \mu_1(\boldsymbol{C}_k)\, \boldsymbol{x}^{\mathsf{T}}\boldsymbol{B}^{\mathsf{T}}\boldsymbol{B}\boldsymbol{x} \geqq k\, \|\boldsymbol{B}\boldsymbol{x}\|^2 \tag{7}$$

holds whenever $\boldsymbol{x} \in R^n$.

By virtue of the inequalities (6) and (7) it is clear

$$\boldsymbol{x}_k^{\mathsf{T}}\boldsymbol{A}\boldsymbol{x}_k + k\, \|\boldsymbol{B}\boldsymbol{x}_k\|^2 \leqq \boldsymbol{x}_k^{\mathsf{T}}\boldsymbol{A}\boldsymbol{x}_k + \boldsymbol{x}_k^{\mathsf{T}}\boldsymbol{B}^{\mathsf{T}}\boldsymbol{C}_k\boldsymbol{B}\boldsymbol{x}_k < \frac{1}{k}. \tag{8}$$

Without loss of generality, we may assume that $\lim_{k\to+\infty} \boldsymbol{x}_k = \boldsymbol{x}_0$ and $\boldsymbol{x}_0 \in S$. The functions $\boldsymbol{x}^{\mathsf{T}}\boldsymbol{A}\boldsymbol{x}$ and $\|\boldsymbol{B}\boldsymbol{x}\|$ are continuous on $S$. From (8) we have

$$\boldsymbol{B}\boldsymbol{x}_0 = 0 \quad \text{and} \quad \boldsymbol{x}_0^{\mathsf{T}}\boldsymbol{A}\boldsymbol{x}_0 \leqq 0\,,$$

contrary to our hypotheses. ■

**Theorem 2.2:** *Let the assumption* (B) *be satisfied. Then there exists a constant* $c^* \geqq 0$ *such that for every* $\boldsymbol{A} \in H(c^*)$ *the matrix* $\nabla^2_{\boldsymbol{xx}} L(\boldsymbol{x}^*, \boldsymbol{\lambda}^*, \boldsymbol{A})$ *is positive definite.*

Proof: By the straightforward differentiation we obtain

$$\nabla^2_{\boldsymbol{xx}} L(\boldsymbol{x}^*, \boldsymbol{\lambda}^*, \boldsymbol{A}) = \nabla^2_{\boldsymbol{xx}} F(\boldsymbol{x}^*, \boldsymbol{\lambda}^*) + \nabla \boldsymbol{g}(\boldsymbol{x}^*)\ \nabla^2_{\boldsymbol{yy}} P(0, \boldsymbol{A})\ \nabla \boldsymbol{g}(\boldsymbol{x}^*)^{\mathsf{T}}\,.$$

By using the previous lemma and assumption (B) we obtain immediately the required result. ■

**Theorem 2.3:** *Let the assumption* (B) *be satisfied, let* $V \subset H(\boldsymbol{c}^*)$, *V is supposed to be a compact set. Then there exist* $\varepsilon^* > 0$, $\delta^* > 0$ *such that for all* $\boldsymbol{\lambda} \in U(\boldsymbol{\lambda}^*, \delta^*)$ *and all* $\boldsymbol{A} \in V$, *the problem*

$$\begin{aligned} &\text{minimize } L(\boldsymbol{x}, \boldsymbol{\lambda}, \boldsymbol{A}) \\ &\text{subject to } \boldsymbol{x} \in U(\boldsymbol{x}^*, \varepsilon^*) \end{aligned}$$

*has a unique solution denoted by* $\boldsymbol{x}(\boldsymbol{\lambda}, \boldsymbol{A})$. *Furthermore, for every* $\varepsilon$ *with* $0 < \varepsilon \leqq \varepsilon^*$, *there exists a* $\delta$ *with* $0 < \delta \leqq \delta^*$ *such that*

$$\boldsymbol{x}(\boldsymbol{\lambda}, \boldsymbol{A}) \in U(\boldsymbol{x}^*, \varepsilon) \quad \textit{for all} \quad \boldsymbol{\lambda} \in U(\boldsymbol{\lambda}^*, \delta) \quad \textit{and} \quad \boldsymbol{A} \in V\,.$$

**Remark 2.4:** The proof of the previous theorem depends heavily on the implicit function theorem. The complete proof of a slightly modified theorem could be found in [2]. The theorem 2.3. essentially says that by locally minimizing the augmented Lagrangian, we obtain points which are arbitrarily and uniformly close to $\boldsymbol{x}^*$ provided $\boldsymbol{\lambda}$ is sufficiently close to $\boldsymbol{\lambda}^*$.

The proofs of the following two lemmas are omitted, see e.g. [6].

**Lemma 2.5:** *Let* $\boldsymbol{A}, \boldsymbol{B}, \boldsymbol{C}, \boldsymbol{D}$ *be symmetric* $m \times m$*-matrices and let* $\boldsymbol{C}, \boldsymbol{D}$ *be positive definite*, i.e. $\boldsymbol{C}, \boldsymbol{D} \in H^0(0)$. *Then*

$$\begin{aligned} &\mu_1(\boldsymbol{A} + \boldsymbol{B}) \geqq \mu_1(\boldsymbol{A}) + \mu_1(\boldsymbol{B})\,, \\ &\mu_m(\boldsymbol{A} + \boldsymbol{B}) \leqq \mu_m(\boldsymbol{A}) + \mu_m(\boldsymbol{B})\,, \\ &\mu_1(\boldsymbol{C}\boldsymbol{D}) \geqq \mu_1(\boldsymbol{C})\, \mu_1(\boldsymbol{D})\,, \\ &\mu_m(\boldsymbol{C}\boldsymbol{D}) \leqq \mu_m(\boldsymbol{C})\, \mu_m(\boldsymbol{D})\,. \end{aligned}$$

**Lemma 2.6:** *Let $A$ be a regular $n\times n$-matrix, $U, V$ be $n\times m$-matrices, $m\leqq n$. Then the matrix $A+UV^{\mathsf{T}}$ is regular if and only if $I+V^{\mathsf{T}}A^{-1}U$ is regular and the following equality holds:*

$$(A+UV^{\mathsf{T}})^{-1}=A^{-1}-A^{-1}U\,(I+V^{\mathsf{T}}A^{-1}U)^{-1}\,V^{\mathsf{T}}A^{-1}\,.$$

## 3. Method of Multipliers

**Definition 3.1:** Let $M^P$, $M^S$ be penalty update formulas. Then by the *method of multipliers for problem* (1) *with $M^P$ and $M^S$*, we mean the following algorithm:

Step 1: Determine $\lambda_1\in R^m$, $A_1^P$, $A_1^S\in G$ set $k:=1$.

Step 2: Find $x_k$ such that

$$L(x_k, \lambda_k, A_k^P)=\min_x L(x, \lambda_k, A_k^P)\,. \tag{9}$$

Step 3: Set

$$A_{k+1}^P=M^P(x_k, \lambda_k, A_k^P)\,, \tag{10}$$
$$A_{k+1}^S=M^S(x_k, \lambda_k, A_k^S)\,, \tag{11}$$
$$\lambda_{k+1}=\lambda_k+\nabla_y P(g(x_k), A_k^S)\,. \tag{12}$$

Step 4: Go to step 2 with $k:=k+1$.

**Example 3.2:** Setting $G=R^1$ and $P(y, A)=\frac{1}{2}A\,\|y\|^2$ for $A\in R^1$, $M^P=M^S$, $M^P$, $M^S$- identical mappings not dependent on $x$ and $\lambda$, we obtain the original method of multipliers discussed in [4], [7].

## 4. Local Convergence Results

In fact, we are interested in iterative procedures which depend on a parameter, i.e. iterations of the form

$$\lambda_{k+1}=S(\lambda_k, A_k), \quad k=1, 2, \ldots, \tag{14}$$

where $S$ is an operator, $S: R^q\times G\to R^q$, $G$ is a metric space.

**Definition 4.1:** Suppose that $V$ is a subset of a metric space $G$, $\lambda^*\in R^q$. The point $\lambda^*$ is said to be a *fixed point of $S$ in* (14) *relative to $V$*, if

$$\lambda^*=S(\lambda^*, A) \tag{15}$$

for all $A\in V$.

The iterative procedure (14) *is locally convergent to $\lambda^*$ with respect to $V$*, if there exists a neighborhood $U(\lambda^*)$ such that for any $\lambda_1\in U(\lambda^*)$, the iteration (14) with any $A_k\in V$ converges to $\lambda^*$. For a point $\lambda^*$ of $R^q$ we set

$$\alpha(V)=\sup\,\{\|\nabla_\lambda S(\lambda^*, A)\|\mid A\in V\}\,. \tag{16}$$

**Theorem 4.2:** *Let $V \subset G$, $V$ be compact, $U \subset R^q$, $U$ be open, $\boldsymbol{\lambda}^* \in U$ be a fixed point of $S$ relative to $V$, $\nabla_\lambda S$ be continuous on $U \times G$, $\alpha(V) < 1$. Then the procedure* (14) *is locally convergent to $\boldsymbol{\lambda}^*$ with respect to $V$.*

Proof: By the compactness of $V$ the function $\varphi$ defined by the following relation

$$\varphi(\boldsymbol{\lambda}) = \sup \{\| \nabla_\lambda S(\boldsymbol{\lambda}, \boldsymbol{A})\| \mid \boldsymbol{A} \in V\}$$

is continuous on $U$.

By this fact and by (14) and (15) there exists a constant $w$ with $0 < w < 1$ and a neighborhood $U(\boldsymbol{\lambda}^*) \subset U$ of $\boldsymbol{\lambda}^*$ such that

$$\|\boldsymbol{\lambda}_{k+1} - \boldsymbol{\lambda}^*\| \leqq \| \nabla_\lambda S\,(\boldsymbol{\lambda}^* + v_k\,(\boldsymbol{\lambda}_k - \boldsymbol{\lambda}^*),\, \boldsymbol{A}_k)\|\, \|\boldsymbol{\lambda}_k - \boldsymbol{\lambda}^*\| \leqq w\, \|\boldsymbol{\lambda}_k - \boldsymbol{\lambda}^*\| \tag{17}$$

for all $k = 1, 2, \ldots$, where $v_k \in [0, 1]$, provided that $\boldsymbol{\lambda}_1 \in U(\boldsymbol{\lambda}^*)$. The local convergence of (14) follows directly from (17). ∎

**Remark 4.3:** When we are able to demonstrate the local convergence of (14) with respect to $V$, we automatically have a local convergence theorem for the iterative method using any update formula $M$ with the only property that there exist a neighborhood $U$ of $\boldsymbol{\lambda}^*$ such that

$$M(U, V) \subset V.$$

No assumptions concerning continuity or differentiability of $M$ are needed.

When we say that the *method of multipliers is locally convergent with respect to $V$*, we mean the general algorithm with unspecified penalty update formulas.

**Theorem 4.4:** *Let the assumption* (B) *be satisfied and let $\boldsymbol{A}^* \in G$ be a given parameter with $\boldsymbol{A}^* \in H(c^*)$. For $\boldsymbol{A}^1, \boldsymbol{A}^2 \in G$ denote*

$$\boldsymbol{M}^1 = \nabla^2_{\boldsymbol{yy}} P(0, \boldsymbol{A}^1)\,,$$
$$\boldsymbol{M}^2 = \nabla^2_{\boldsymbol{yy}} P(0, \boldsymbol{A}^2)\,,$$
$$\boldsymbol{M}^* = \nabla^2_{\boldsymbol{yy}} P(0, \boldsymbol{A}^*)\,.$$

*Define*

$$H = \{(\boldsymbol{A}^1, \boldsymbol{A}^2) \in G \times G \mid 0 < \mu_1(\boldsymbol{M}^1),\, c^* \leqq \mu_1(\boldsymbol{M}^2),\, \mu_m(\boldsymbol{M}^1) \leqq 2\mu_1\,(\boldsymbol{M}^2 - \boldsymbol{M}^*)\}$$

*and let $V \subset H$, $V$ be compact, $c^*$ is from the theorem 2.2. Then the method of multipliers is locally convergent to $\boldsymbol{\lambda}^*$ with respect to $V$. Moreover,* $\lim \boldsymbol{x}_k = \boldsymbol{x}^*$, *where* $\boldsymbol{x}_k = \boldsymbol{x}(\boldsymbol{\lambda}_k, \boldsymbol{A}_k^2)$.

Proof: To prove the theorem we use the theorems 2.3. and 4.2. The iteration mapping has the following form

$$S(\boldsymbol{\lambda}, (\boldsymbol{A}^1, \boldsymbol{A}^2)) = \boldsymbol{\lambda} + \nabla_{\boldsymbol{y}} P(\boldsymbol{g}(\boldsymbol{x}(\boldsymbol{\lambda}, \boldsymbol{A}^2)), \boldsymbol{A}^1)\,. \tag{18}$$

A straightforward differentiation shows that

$$\nabla_\lambda S(\boldsymbol{\lambda}^*, (\boldsymbol{A}^1, \boldsymbol{A}^2)) = \boldsymbol{I} - \boldsymbol{M}^1 \nabla \boldsymbol{g}^{*\mathsf{T}} (\nabla^2_{\boldsymbol{xx}} L^* + \nabla \boldsymbol{g}^*\,(\boldsymbol{M}^2 - \boldsymbol{M}^*) \nabla \boldsymbol{g}^{*\mathsf{T}})^{-1} \nabla \boldsymbol{g}^*$$

where

$$\boldsymbol{g}^* = \boldsymbol{g}(\boldsymbol{x}^*), \quad L^* = L(\boldsymbol{x}^*, \boldsymbol{\lambda}^*, \boldsymbol{A}^*)\,. \tag{19}$$

Denote

$$A^{**} = \nabla g^{*\mathrm{T}}(\nabla^2_{xx}L^*)^{-1}\nabla g^* . \tag{20}$$

By the lemma 2.5. and by (20) we obtain

$$(M^1)^{-1/2}\nabla_\lambda S(M^1)^{1/2} = I - (M^1)^{1/2}[(A^{**})^{-1} + (M^2 - M^*)]^{-1}(M^1)^{1/2} .$$

It follows from the lemma 2.6. that

$$\mu_1(\nabla_\lambda S) \geqq 1 - \frac{\mu_m(M^1)\,\mu_m(A^{**})}{1 + \mu_1(M^2 - M^*)\,\mu_m(A^{**})} ,$$

$$\mu_m(\nabla_\lambda S) \leqq 1 - \frac{\mu_1(M^1)\,\mu_1(A^{**})}{1 + \mu_m(M^2 - M^*)\,\mu_1(A^{**})} .$$

For the spectral radius $\varrho(\nabla_\lambda S)$ of the matrix $\nabla_\lambda S(\lambda^*, (A^1, A^2))$ we have

$$\varrho(\nabla_\lambda S) \leqq \max\left\{\left|1 - \frac{\mu_m(M^1)\,\mu_m(A^{**})}{1 + \mu_1(M^2 - M^*)\,\mu_m(A^{**})}\right| , \left|1 - \frac{\mu_1(M^1)\,\mu_1(A^{**})}{1 + \mu_m(M^2 - M^*)\,\mu_1(A^{**})}\right|\right\} . \tag{21}$$

The right hand side of the inequality (21) is less then unity if and only if

$$\mu_m(M^1) - 2\mu_1(M^2 - M^*) < \frac{2}{\mu_m(A^{**})} \tag{22}$$

and

$$\mu_1(M^1) - 2\mu_m(M^2 - M^*) < \frac{2}{\mu_1(A^{**})} . \tag{23}$$

Consequently, by hypotheses of the theorem both inequalities (22) and (23) are satisfied and, accordingly, for every pair $(A^1, A^2) \in V$ it follows that $\varrho(\nabla_\lambda S) \leqq \varepsilon < 1$. The rest follows from the theorem 4.2. The last part of the theorem follows from the theorem 2.3. ■

**Remark 4.5:** The previous theorem needn't comprise special and important cases of the penalty update formula: $M^P = M^S$ and $M^P = M^S + A^*$ where $A^*$ is a preselected penalty parameter with $A^* \in H(c^*)$ where $M^i$: $R^n \times R^m \times G \to G$, $i \in \{p, s\}$, $G$ is a metric linear space. In these cases, however, convergence theorems, similar to the theorem 4.4., can be formulated and proved by the similar manner as in the case of the theorem 4.4.

**Theorem 4.6:** *Let the assumption* (B) *be satisfied and $A^*$ be a given penalty parameter with $A^* \in H(c^*)$, let $V \subset H^0(0)$ and $V$ be a compact set. Then the method of multipliers using the penalty update formula $M^P = M^S + A^*$ is locally convergent with respect to $V$.*

Proof: The method of multipliers is expressed by the iteration mapping

$$S(\lambda, A) = \lambda + \nabla_y P(g(x(\lambda, A + A^*)), A) .$$

Now, the proof of the theorem is similar to the proof of the theorem 4.4.

**Theorem 4.7:** *Let the assumption* (B) *be satisfied, let $\alpha \geqq 2$ and define*

$$H^\alpha = \{A \in G;\ \mu_1(\nabla^2_{yy}P(0, A)) \geqq \alpha c^*,\ \mu_m(\nabla^2_{yy}P(0, A)) \leqq 2\,(\alpha - 1)\,c^*\} .$$

*Let $V \subset H^\alpha$, $V$ be a compact set.*

*Then the method of multipliers using the penalty update formula $M^P = M^S$ is locally convergent with respect to $V$.*

Proof: Let $M^P = M^S$. Consider the following iteration operator

$$S(\boldsymbol{\lambda}, \boldsymbol{A}) = \boldsymbol{\lambda} + \nabla_y P(g(\boldsymbol{x}(\boldsymbol{\lambda}, \boldsymbol{A}), \boldsymbol{A}) ,$$

where $\boldsymbol{A} \in V$. This mapping is investigated by the same manner as in the previous theorems.

**Remark 4.8:** The previous convergence results are proved under the assumption that $\boldsymbol{\lambda}_1$ — the first approximation of the LAGRANGE multiplier $\boldsymbol{\lambda}$ is sufficiently close to the optimal LAGRANGE mutiplier $\boldsymbol{\lambda}^*$. The author of this paper has the feeling that by adding suitable hypotheses on $f$, $\boldsymbol{g}$ and $P$, e.g. LIPSCHITZ connuity of the HESSIAN matrices of $F$ and $P$ then for an arbitrary $\boldsymbol{\lambda}_1$ there exists a $c^{**}$ sufficiently large such that $\boldsymbol{\lambda}_2$ lies appropriately close to $\boldsymbol{\lambda}^*$ and the method converges. This was proved by BERTSEKAS for the method of the example 3.2., see [3].

## 5. Concluding Remarks

In this paper, the method of multipliers for obtaining local convergence was investigated. The usual penalty constant was replaced by a more general penalty parameter, the method of multipliers was generalized. Further, the sufficient conditions for local convergence of this method were formulated. The convergence analysis reveals a connection between the eigenvalue structure of the HESSIAN matrix of the penalty function at the solution point and the local convergence properties of the method.

As concerns a rate of convergence of our method we can state that the rate is at least linear. This fact follows from the corresponding proofs.

The presented forms of the augmented LAGRANGians and wide variety of penalty parameters, e.g. in a matrix form, give new computational possibilities for solving problems with equalities. Due to this fact the given convergence analysis may be useful.

### References

[1] ARROW, K. J.; GOULD, F. J.; HOWE, S. M.: A general saddle point result for constrained optimization. Math. Progr. **5** (1973) 225–234.

[2] BERTSEKAS, D. P.: Combined primal-dual and penalty methods for constr. minimization. SIAM J. on Contr. **13** (1975) 521–544.

[3] BERTSEKAS, D. P.: On penalty and multiplier methods for constr. minimization. SIAM J. on Contr. **14** (1976) 2.

[4] HESTENES, M. R.: Multiplier and gradient methods. JOTA **4** (1969) 303–320.

[5] MANGASARIAN, O. L.: Unconstrained Lagrangians in nonlin. programming **13** (1975) 772–791.

[6] ORTEGA, J. M.-RHEINBOLDT, W. C.: Iterative solution of nonlin. equations in several variables. Acad. Press, New York 1970.
[7] POWELL, M. J.: A method for nonlin. constr. in minimization problems. In: Optimization (ed. R. FLETCHER) Acad. Press, New York 1969, pp. 283–298
[8] ROCKAFELLAR, R. T.: Augmented Lagrange multiplier functions and duality in nonconvex programming. SIAM J. on Contr. 12 (1974) 268–285.
[9] ROCKAFELLAR, R. T.: Solving a nonlin. progr. problem by the way of dual problem Symposia Mathematica 19 (1976) 135–160.
[10] ROCKAFELLAR, R. T.: Lagrange multipliers in optimization. SIAM-AMS Proc. 9 (1976) 145–168.
[11] TAPIA, R. A.: Diagonalized mult. methods . . ., JOTA 22 (1977) 135–194.

**Zusammenfassung**

In dieser Arbeit befassen wir uns mit der Methode der Multiplikatoren von HESTENES-POWELL zur Lösung von Minimierungsproblemen mit Gleichungsrestriktionen.

Es werden verallgemeinerte LAGRANGE-Funktionen benutzt, die nicht notwendig konkav in dem verallgemeinerten Strafparameter sind. Es werden lokale Konvergenzaussagen bewiesen.

**Резюме**

В этой статье мы занимаемся обобщенным методом ГЭСТЕНСА-ПАУЭЛЛА, тоже называемым методом множителей, для решения экстремальных задач с ограничениями. Введенный расширенный Лагранжиан не обязательно вогнутый во взгляду на штрафный параметр. Этот параметр тоже обобщается. Только ограничения типа равенств рассматриваются в статье.

Received December 1979; revised January 1980.

## Book Review

J. K. Lenstra: **Sequencing by Enumerative Methods.** Mathematical Centre Tracts 69. Mathematisch Centrum Amsterdam 1977, 198 S.

Die aus der Dissertation des Autors entstandene Studie zur Reihenfolgeproblematik wendet sich an diejenigen Mathematiker, Rechentechniker, Ingenieure und Ökonomen (bzw. Studierende entsprechender Fachrichtungen), die auf Gebieten arbeiten, wo kombinatorische Optimierungsprobleme eine Rolle spielen.

Behandelt werden im wesentlichen zwei Klassen kombinatorischer Optimierungsprobleme, von welchen die eine das quadratische Zuordnungsproblem und seine verschiedenen Modifikationen wie das azyklische Teilgraphenproblem und das klassische Problem des Handlungsreisenden (Rundfahrtproblem) enthält und die andere, größere, aus Maschinen-Reihefolgeproblemen besteht, welche immer dann entstehen, wenn Aufträge durch kapazitätsbeschränkte Maschinen zu bearbeiten sind.

Im Teil I wird die Komplexität verschiedener Reihenfolgeprobleme aus beiden Klassen studiert. Die Ergebnisse gestatten, eine detaillierte Lokalisation der imaginären „Grenzlinie" zwischen „leichten" und „schweren" Reihefolgeproblemen (unter der $P \neq NP$ Hypothese). Insbesondere ergibt sich, daß die meisten, praktisch interessanten Probleme Lösungsprozeduren erfordern, die auf der Enumeration der zulässigen Lösungen basieren bzw. approximativen oder heuristischen Charakters sind.

Teil II ist den enumerativen Methoden gewidmet, deren Effektivität bekanntermaßen von der gewählten Rechner-Implementation bestimmt wird. Der Autor trägt dem dadurch Rechnung, daß er einfaches, wirkungsvolles, rekursives Vorgehen für die Implementierung entwickelt, welches auf einfach strukturierte Schemata für die explizite und implizite Enumeration führt.

Der III. Teil wendet sich an die Lösung von Reihenfolgeproblemen mittels impliziter Enumeration. Für das Rundfahrtproblem sowie zwei 1-Maschinen- und zwei $m$-Maschinen-Reihenfolgeprobleme werden zum Teil bekannte Branch- and Bound-Prozeduren in ALGOL bzw. quasi-ALGOL beschrieben und ihr rechnerisches Verhalten studiert.

Anwendungen der Theorie von Reihenfolgeproblemen stehen im Mittelpunkt des vierten Teils. Gezeigt wird, wie sich Probleme der Verdrahtung in Rechnern, der Fahrtroutenbestimmung von Kraftfahrzeugen sowie der Werkstattfertigung von Aufträgen ohne Wartezeiten erfolgreich als Rundfahrtproblem lösen lassen.

Das Literaturverzeichnis umfaßt über 250 Artikel, darunter jedoch keine aus der DDR, bedauerlicherweise muß man sagen, denn insbesondere für die Teile III und IV hätte der Autor davon profitieren können.

Vom Leser wird erwartet, daß er mit den Grundlagen der Mathematischen Optimierung, der Graphentheorie sowie der Programmierung vertraut ist.

B. Bank

Math. Operationsforsch. Statist., Ser. Optimization, Vol. 11 (1980) No. 4, 547–561

# Zur mathematischen Theorie der optimalen Standardisierung[1]

E. GIRLICH[2] M. M. KOWALJOW[3]

**Zusammenfassung:** In dieser Arbeit werden Probleme der optimalen Standardisierung untersucht. Es werden Ergebnisse zur Struktur des zulässigen Lösungsbereiches dargelegt. Diese bilden die Grundlage für Lösungsmethoden dieser Aufgabenklasse.

## 0. Einleitung

In dieser Arbeit werden die Untersuchungen zum Probem der optimalen Standardisierung [1, 2] (POS) fortgesetzt. Die Strukturuntersuchungen der zulässigen Lösungsbereiche stehen im Vordergrund. Analoge Untersuchungen sind zu den Polyedern linearer Optimierungsprobleme [5, 6], des Transportproblems [7], des Rundreiseproblems [8] angestellt worden. Die bisherigen Resultate wurden zur Konstruktion kombinatorischer [1] und lokaloptimaler stochastischer [3] Lösungsverfahren verwendet.

Hier wird nachgewiesen, daß die zulässigen Lösungsbereiche des POS beschränkte Polymatroide, eine spezielle von EDMONDS [4] betrachtete Klasse von Polyedern, unbeschränkte Polymatroide bzw. der Durchschnitt von Polymatroiden mit einer Hyperebene sind.

Eigenschaften der Polymatroide gestatteten die Konstruktion von Lösungsalgorithmen des konvexen POS [11].

## 1. Äquivalente Darstellungen des Polyeders $H(Q, b)$

In [1] ist eine inhaltliche Darstellung und Klassifizierung der POS gegeben, die zu folgendem mathematischen Modell führen:

$$F(Y)=\sum_{i=1}^{m} F_i(y_i)\to\min \tag{1}$$

$$\sum_{j=1}^{n} x_{ij}=y_i, \quad i=1, \ldots, m, \tag{2}$$

[1] Resultate wurden vorgetragen auf den Tagungen „Mathematische Optimierung – Theorie und Anwendungen", Eisenach 1977 und 1978, DDR.

[2] FSU Jena, Sektion Mathematik, DDR-69 Jena.

[3] BGU Minsk, Fakultät Angewandte Mathematik, 220080 Minsk, UdSSR.

$$\sum_{i=1}^{m} q_{ij} x_{ij} = b_j, \quad j=1, \ldots, n, \tag{3}$$

$$x_{ij} \begin{cases} \geqq 0 & q_{ij} > 0, \\ = 0 & q_{ij} = 0, \end{cases} \quad i=1, \ldots, m, \quad j=1, \ldots, n, \tag{4}$$

wobei wir o. B. d. A. annehmen können, daß für alle $j$ gilt $b_j > 0$ und ganzzahlig sowie $q_{ij} = 0 \vee 1$, $i=1, \ldots, m$, $j=1, \ldots, n$ und $\sum_{i=1}^{m} q_{ij} \geqq 2$. Aus ökonomischen Überlegungen folgt, daß die Produktionskostenfunktionen $F_i(y_i)$ als konkav angenommen werden können. Daher ist das Minimum auf dem Rand, in den Ecken des Polyeders $H(Q, b)$, $H(Q, b) = \left\{ y \in E^m \mid y_i = \sum_{j=1}^{n} x_{ij}, \ x \in G(Q, b) \right\}$, wobei $G(Q, b)$ beschrieben wird durch die Bedingungen (3), (4), zu finden.

Über einen Satz von Gale [9] über die Existenz eines Stromes über einem entsprechenden paaren Graphen kommen wir zu einer direkten Darstellung des Polyeders $H(Q, b)$ durch (5)–(7)

$$\sum_{i \in w(J)} y_i \geqq \sum_{j \in J} b_j, \quad J \neq \emptyset, \quad J \subsetneqq \{1, \ldots, n\}, \tag{5}$$

$$y_i \geqq 0, \quad i=1, \ldots, m, \tag{6}$$

$$\sum_{i=1}^{m} y_i = \sum_{j=1}^{n} b_j, \tag{7}$$

wobei $w(J) = \{i \mid \sum_{j \in J} q_{ij} \geqq 1\}$ ist.

Ersetzen wir (5) durch (5')

$$\sum_{i \in w(J)} y_i \geqq \sum_{j \in J} b_j, \quad J \neq \emptyset, \quad J \subseteqq \{1, \ldots, n\}, \tag{5'}$$

wobei $w(J)$ wie bei (5) definiert ist, so ist mit (5'), (6) ein Polyeder $H'(Q, b)$ beschrieben.

Die Bedingungen (5)–(7) ((5'), (6)) enthalten $2^n + m - 2$ $(2^n + m - 1)$ Ungleichungen. Leider ist dieses System für die uns interessierenden Fälle $n \geqq m$ meist redundant. Betrachten wir folgendes Ungleichungssystem

$$\sum_{i \in I} y_i \geqq \sum_{j \in v(I)} b_j, \quad I \neq \emptyset, \quad I \subsetneqq \{1, \ldots, m\}, \tag{8}$$

wobei $v(I) = \{j \mid \sum_{i \notin I} q_{ij} = 0\}$ ist.

**Satz 1:** *Die Darstellungen des Polyeders* $H(Q, b)$ *durch* (5)–(7) *und* (7), (8) *sind äquivalent.*

Beweis: (7) ist in beiden Darstellungen erfüllt. Wir zeigen zuerst: Aus (8) folgt (5), (6).

a) Für einelementige Mengen $I = \{i_1\}$ erhalten wir mit der Voraussetzung $\sum_{i=1}^{m} q_{ij} \geqq 2$, daß $v(I)=\emptyset$ und damit gilt

$$\sum_{i\in I=\{i_1\}} y_i = y_{i_1} \geqq \sum_{j\in v(I)} b_j = \sum_{j\in\emptyset} b_j = 0, \quad \text{also} \quad y_{i_1} \geqq 0\,,$$

somit gilt (6).

b) Sei nun $|I|\geqq 2$. Nehmen wir an, es existiere eine Menge $J_0$, für welche (5) nicht erfüllt ist. Dann können wir mit $I_0=w(J_0)$ eine Menge betrachten, für die (8) gilt. Weisen wir nach, daß $v(I_0)\supset J_0$, so erhalten wir aus

$$\sum_{j\in J_0} b_j > \sum_{i\in w(J_0)} y_i = \sum_{i\in I_0} y_i \geqq \sum_{j\in v(I_0)} b_j \geqq \sum_{j\in J_0} b_j$$

einen Widerspruch zur Voraussetzung, daß für $J_0$ (5) nicht gilt. Sei $j_0\in J_0$, dann können wir mit $I_{j_0}=\{i\mid q_{ij_0}=1\}$ bezeichnen. Es gilt $I_{j_0}\subset I_0=w(J_0)$. Dann muß $j_0\in v(I_{j_0})=\{j\mid \sum_{i\notin I_{j_0}} q_{ij}=0\}$ gelten, und somit, da $I_{j_0}\subset I_0, j_0\in v(I_0)$. Also gilt $J_0\subset v(I_0)$.

Wir zeigen nun die entgegengesetzte Richtung. Nehmen wir an, für $I_0$ sei (8) nicht erfüllt, aber für $J_0=v(I_0)\neq\emptyset$ sei (5) erfüllt. Wenn wir zeigen, daß $w(J_0)\subset I_0$ gilt, dann erhalten wir mit

$$\sum_{i\in I_0} y_i < \sum_{j\in v(I_0)} b_j = \sum_{j\in J_0} b_j \leqq \sum_{i\in w(J_0)} y_i \leqq \sum_{i\in I_0} y_i$$

einen Widerspruch zur Voraussetzung, daß für $I_0$ (8) nicht erfüllt ist. Sei $i_0\in w(J_0)= =\{i\mid \sum_{i\in J_0} q_{ij}\geqq 1\}$, dann existiert mindestens ein $j_0\in J_0$ mit $q_{i_0j_0}=1$. Da $J_0=v(I_0)$ ist, gilt $j_0\in J_0=\{i\mid \sum_{i\notin I_0} q_{ij}=0\}$, also muß $i_0\in I_0$ sein. Folglich gilt also $w(J_0)\subset I_0$, für $v(I_0)\neq\emptyset$. Sei nun $J_0=v(I_0)=\emptyset$. Dann erhalten wir aus (6) bei der Annahme, daß $I_0$ (8) nicht erfüllt, den Widerspruch

$$0=\sum_{j\in\emptyset} b_j > \sum_{i\in I_0} y_i \geqq \sum_{j\in J_0} b_j = 0\,.$$

Damit ist Satz 1 bewiesen. ∎

**Folgerung 1:** Fügen wir diesem Beweis die Äquivalenz der Ungleichungen

$$\sum_{i\in w(J')} y_i \geqq \sum_{j\in J'} b_j\,, \quad J'=\{1,\ldots,n\}$$

und

$$\sum_{i\in I'} y_i \geqq \sum_{j\in v(I')} b_j\,, \quad I'=\{1,\ldots,m\}$$

an, was auf Grund von $w(J')=\{1,\ldots,m\}$, $v(I')=\{1,\ldots,n\}$ gilt, so erhalten wir mit (8′) eine Darstellung des Polyeders $H'(Q,b)$. (8′) unterscheidet sich von (8) durch Hinzunahme der Ungleichung mit $I=\{1,\ldots,n\}$.

**Folgerung 2:** Das Polyeder $H(Q,b)$ ergibt sich als Durchschnitt des Polyeders $H'(Q,b)$ mit der Hyperebene $\sum_{i=1}^{m} y_i = \sum_{j=1}^{n} b_j$.

Ausgehend vom System (7), (8) erhalten wir eine weitere Darstellung des Polyeders $H(Q, b)$ durch (7), (9)

$$\sum_{i\in I} y_i \leqq \sum_{j\in u(I)} b_j, \quad I \neq \emptyset, \quad I \subsetneqq \{1, \ldots, m\}, \tag{9}$$

wobei $u(I)=\{j \mid \sum_{i\in I} q_{ij} \geqq 1\}$ ist.

**Satz 2:** *Die Darstellungen des Polyeders $H(Q, b)$ durch die Bedingungen* (7), (8) *und* (7), (9) *sind äquivalent.*

Beweis: Die Bedingung (7) ist in beiden Systemen vorhanden. Sei (8) erfüllt für alle $I \subset \{1, \ldots, m\}$; wir zeigen, daß dann auch (9) erfüllt wird. Sei $I_0$ eine beliebige nichtleere Teilmenge von $\{1, \ldots, m\}$, dann hat $\bar{I}_0=\{1, \ldots, m\} \setminus I_0$ die Eigenschaft $I_0 \cup \bar{I}_0=\{1, \ldots, m\}$, $I_0 \cap \bar{I}_0=\emptyset$. Für $\bar{I}_0$ gelte (8), und wir erhalten mit (7)

$$\sum_{i=1}^{m} y_i - \sum_{i\in \bar{I}_0} y_i \leqq \sum_{j=1}^{n} b_j - \sum_{j\in v(\bar{I}_0)} b_j,$$

was gleichbedeutend ist mit

$$\sum_{i\in I_0} y_i \leqq \sum_{j=1}^{n} b_j - \sum_{j\in v(\bar{I}_0)} b_j = \sum_{j\in u(I_0)} b_j,$$

da $\{1, \ldots, n\} \setminus v(\bar{I}_0)=\{j \mid j \notin v(\bar{I}_0)\}$

$$=\{j \mid \sum_{i\notin \bar{I}_0} q_{ij} \geqq 1\}$$

$$=\{j \mid \sum_{i\in I_0} q_{ij} \geqq 1\} = u(I_0) \text{ ist}.$$

Es gilt somit für $I_0, \bar{I}_0, I_0 \cup \bar{I}_0=\{1, \ldots, m\}, I_0 \cap \bar{I}_0=\emptyset$ und $u(I_0) \cup v(\bar{I}_0)=\{1, \ldots, n\}$. Damit ist diese Richtung bewiesen, der Beweis der anderen Richtung erfolgt analog. ∎

Mit $H''(Q, b)$ wollen wir folgendes Polyeder bezeichnen

$$\{y \in E^{m+} \mid \sum_{i\in I} y_i \leqq \sum_{j\in u(I)} b_j,\ I \neq \emptyset,\ I \subseteqq \{1, \ldots, m\}\}. \tag{9'}$$

**Folgerung 3:** Das Polyeder $H(Q, b)$ ergibt sich als Durchschnitt des Polyeders $H''(Q, b)$ mit der Hyperebene $\sum_{i=1}^{m} y_i = \sum_{j=1}^{n} b_j$, da alle $b_j > 0$ vorausgesetzt sind.

**Folgerung 4:** Das Polyeder $H(Q, b)$ ergibt sich als Durchschnitt der Polyeder $H'(Q, b)$ und $H''(Q, b)$.

Aus allgemeinen Aussagen der Polyederdarstellung [10] erhalten wir

**Folgerung 5:**

$$\dim H'(Q, b) = \dim H''(Q, b) = m,$$
$$\dim H(Q, b) = m-1.$$

**Folgerung 6:** Die Anzahl $f_{m-2}$ der $(m-2)$-dimensionalen Seiten (Facetten) des Polyeders $H(Q, b)$ übersteigt $2^m-2$ nicht.

**Folgerung 7:** Die Facettenanzahl $f_{m-2}(H(Q, b)) = f_{m-1}(H''(Q, b))$ ist durch $2^m-1$ nach oben beschränkt.

## 2. Polymatroide

In diesem Abschnitt wollen wir den Begriff des Polymatroiden [4] und unbeschränkten Polymatroiden [11] einführen.

**Definition 1:** Eine Funktion $R(I)$, definiert auf der Potenzmenge der Menge $\{1, \ldots, m\}$, besitzt die Eigenschaft $\alpha$, wenn folgende Bedingungen erfüllt sind:

1. $R(\emptyset) = 0$ ,
2. $R(I) \leqq R(I')$ für $I \subset I' \subset \{1, \ldots, m\}$ ,
3. $R(I) + R(I') \geqq R(I \cap I') + R(I \cup I')$, $I, I' \subset \{1, \ldots, m\}$ .

**Definition 2:** Eine Funktion $R(I)$, definiert auf der Potenzmenge der Menge $\{1, \ldots, m\}$, besitzt die Eigenschaft $\beta$, wenn die Beziehungen 1. und 2. (wie bei Definition 1) und 3. $R(I) + R(I') \leqq R(I \cap I') + R(I \cup I')$, $I, I' \subset \{1, \ldots, m\}$ erfüllt sind.

**Definition 3:** Das Polyeder $P(R)$ im $E^{m+}(E^m)$, gegeben durch die Restriktionen

$$\sum_{i \in I} y_i \underset{(\geqq)}{\leqq} R(I) \quad \text{für alle} \quad I \subset \{1, \ldots, m\} ,$$

nennen wir *Polymatroid* (*unbeschränktes Polymatroid*), wenn die Funktion $R(I)$ die Eigenschaft $\alpha$ $(\beta)$ besitzt.

**Satz 3:** *Ein Polyeder $P$ im $E^{m+}(E^m)$ ist genau dann Polymatroid (unbeschränktes Polymatroid), wenn es folgende Eigenschaften besitzt:*

1. *Ist $x \in P$ und $y \leqq x$ $(x \leqq y)$, so ist $y \in P$.*
2. *Für alle maximalen (minimalen) Elemente $x'$ der Menge $\{x \in P \mid x \leqq a \; (x \geqq a)\}$ gilt für einen beliebigen Vektor $a \in E^{m+}$ die Beziehung*

$$\sum_{i=1}^{m} x_i' = r(a) ,$$

*wobei $r(a)$ eine von $a$ abhängige Konstante ist.*

Dieser in [4] formulierte Satz für Polymatroide kann in dieser Form auch für die unbeschränkten Polymatroide bewiesen werden.

**Satz 4:** *Das Permutationspolyeder*

$$P(b) = \left\{ x \in E^{m+} \mid \sum_{i \in I} x_i \leqq \sum_{j=1}^{|I|} b_j , \; I \subseteqq \{1, \ldots, m\} \right\}$$

*mit $b_1 > b_2 > \ldots > b_m \geqq 0$ ist ein Polymatroid.*

*Das Permutationspolyeder*

$$P(b') = \left\{ x \in E^m \mid \sum_{i \in I} x_i \geqq \sum_{j=1}^{|I|} b_j' , \; I \subseteqq \{1, \ldots, m\} \right\}$$

*mit $0 \leqq b_1' < \ldots < b_m'$ ist ein unbeschränktes Polymatroid.*

Beweis: Entsprechend Definition 3 ist nachzuweisen, daß die Funktion $R(I)=\sum_{j=1}^{|I|} b_j$ die Eigenschaft $\alpha$ und $R(I)=\sum_{j=1}^{|I|} b'_j$ die Eigenschaft $\beta$ besitzt.

1. $R(\emptyset)=\sum_{j=1}^{|\emptyset|} b_j=0, \quad R(\emptyset)=\sum_{j=1}^{|\emptyset|} b'_j=0.$
2. $R(I)\leqq R(I')$ für $I\subset I'\subset\{1,\ldots,m\}$, gilt, da alle $b_j\geqq 0$, $b'_j\geqq 0$ sind.
3. $R(I)+R(I')\geqq R(I\cap I')+R(I\cup I')$, d. h.

$$\sum_{j=1}^{|I|} b_j+\sum_{j=1}^{|I'|} b_j\geqq \sum_{j=1}^{|I\cap I'|} b_j+\sum_{j=1}^{|I\cup I'|} b_j.$$

Nehmen wir an, $I$ besitze $r$, $I'$ $s$ Elemente, $I\cap I'$ $t$ Elemente. Auf der linken Seite wird demnach über die $r$ und $s$ größten Elemente summiert, auf der rechten Seite über die $r+s-t$ und $t$ größten, damit gilt auch

$$\sum_{j=t+1}^{r} b_j+\sum_{j=t+1}^{s} b_j\geqq \sum_{j=t+1}^{s+r-t} b_j.$$

Da auf beiden Seiten die gleiche Anzahl von Elementen, auf der linken Seite die größeren stehen, ist diese Ungleichung erfüllt. Die Gleichheit gilt bei $r=s=t$.

Für $P(b')$ ergibt sich diese Bedingung (3) analog. Damit ist Satz 4 bewiesen. ▮

**Bemerkung:** In [12] wurde gezeigt, daß der Durchschnitt von $P(b')$ mit der Hyperebene $\sum_{i=1}^{m} x_i=\sum_{j=1}^{m} b'_j$ die konvexe Hülle der $m!$ Punkte $(b'_{s_1}, b'_{s_2},\ldots,b'_{s_m})$ ergibt, wobei $s=(s_1,\ldots,s_m)$ eine Permutation der Zahlen $\{1,\ldots,m\}$ und $0\leqq \leqq b'_1<\ldots<b'_m$ ist.

**Satz 5:** *Das Polyeder $H''(Q,b)$ ist ein Polymatroid. Das Polyeder $H'(Q,b)$ ist ein unbeschränktes Polymatroid.*

Beweis: Wir weisen nach, daß die Funktionen $R(I)=\sum_{j\in u(I)} b_j$ die Eigenschaft $\alpha$ bzw. $R(I)=\sum_{j\in v(I)} b_j$ die Eigenschaft $\beta$ besitzen.

a) $R(I)=\sum_{j\in u(I)} b_j$ besitzt die Eigenschaft $\alpha$, da

1. $R(\emptyset)=\sum_{j\in u(\emptyset)} b_j=0$, weil $u(\emptyset)=\emptyset$ ist.
2. $R(I)\leqq R(I')$ für $I\subset I'\subset\{1,\ldots,m\}$, weil $u(I)\subset u(I')$ gilt.
3. $R(I)+R(I')\geqq R(I\cap I')+R(I\cup I')$ für $I, I'\subset\{1,\ldots,m\}$, d. h.

$$\sum_{j\in u(I)} b_j+\sum_{j\in u(I')} b_j\geqq \sum_{j\in u(I\cap I')} b_j+\sum_{j\in u(I\cup I')} b_j.$$

Dazu verwenden wir, daß

$$u(I\cup I')=\{j\mid \sum_{i\in I\cup I'} q_{ij}\geqq 1\}=\{j\mid \sum_{i\in I} q_{ij}\geqq 1 \vee \sum_{i\in I'} q_{ij}\geqq 1\}=u(I)\cup u(I')$$

und

$$u(I \cap I') = \{j \mid \sum_{i \in I \cap I'} q_{ij} \geqq 1\} \subseteqq \{j \mid \sum_{i \in I} q_{ij} \geqq 1 \wedge \sum_{i \in I'} q_{ij} \geqq 1\} \subseteqq u(I) \cap u(I') \quad (11)$$

gilt. Dann können wir die Summation ändern in

$$\sum_{j \in u(I)} b_j + \sum_{j \in u(I')} b_j = \sum_{j \in u(I) \cap u(I')} b_j + \sum_{j \in u(I) \cup u(I')} b_j ,$$

was nach (11)

$$\geqq \sum_{j \in u(I \cup I')} b_j + \sum_{j \in u(I \cap I')} b_j \quad \text{ergibt} .$$

b) $R(I) = \sum_{j \in v(I)} b_j$ besitzt die Eigenschaft $\beta$.

1. $R(\emptyset) = 0$, da $v(\emptyset) = \emptyset$.
2. $R(I) \subseteqq R(I')$ für $I \subset I' \subset \{1, \ldots, m\}$, da $v(I) \subset v(I')$ gilt.
3. $R(I) + R(I') \leqq R(I \cap I') + R(I \cup I')$.

Dazu verwenden wir, daß

$$v(I \cup I') = \{j \mid \sum_{i \notin I \cup I} q_{ij} = 0\} \supseteqq \{j \mid \sum_{i \notin I} q_{ij} = 0 \vee \sum_{i \notin I'} q_{ij} = 0\} \supseteqq v(I) \cup v(I') \quad (12)$$

und

$$v(I \cap I') = \{j \mid \sum_{i \notin I \cap I'} q_{ij} = 0\} = \{j \mid \sum_{i \notin I} q_{ij} = 0 \wedge \sum_{i \notin I'} q_{ij} = 0\} = v(I) \cap v(I') .$$

Damit ergibt sich

$$\sum_{j \in v(I)} b_j + \sum_{j \in v(I')} b_j = \sum_{j \in v(I) \cup v(I')} b_j + \sum_{j \in v(I) \cap v(I')} b_j ,$$

und mit (12)

$$\leqq \sum_{j \in v(I \cup I')} b_j + \sum_{j \in v(I \cap I')} b_j .$$

Damit ist Satz 5 bewiesen. ▮

## 3. Darstellung und Anzahl der Ecken des Polyeders $H(Q, b)$

In [4] wird für die Lösung einer linearen Optimierungsaufgabe

$$\max_{x \in P(R)} cx$$

über einem Polymatroiden $P(R)$ ein Lösungsverfahren angegeben.

**Satz 6:** *Das Maximum von $cx$ über $P(R) \subset E^{m+}$ wird im Punkt $\mathring{x}$ angenommen, der in folgender Weise bestimmt wird: Sei*

$$c_{j_1} \geqq \ldots \geqq c_{j_k} > 0 \geqq c_{j_{k+1}} \geqq \ldots \geqq c_{j_m} , \quad \text{dann ist}$$
$$\mathring{x}_{j_1} = R(\{j_1\}), \quad \mathring{x}_{j_2} = R(\{j_1, j_2\}) - R(\{j_1\}), \ldots ,$$
$$\mathring{x}_{j_k} = R(\{j_1, \ldots, j_k\}) - R(\{j_1, \ldots, j_{k-1}\}) , \quad (13)$$

*und*

$$\mathring{x}_{j_p} = 0, \quad p = k+1, \ldots, m .$$

**Lemma 1:** *Die Ecken eines Polymatroids $P(R)$ ergeben sich aus allen möglichen Permutationen $(s_1, \ldots, s_k, s_{k+1}, \ldots, s_m)$ der Zahlen $1, \ldots, m$ über die Beziehung* (13).

**Folgerung 8:** Jede Ecke des Polymatroiden $H''(Q,b)$ ergibt sich nach einer Permutation $(s_1, \ldots, s_k, s_{k+1}, \ldots, s_m)$ und einer Zahl $k$ über die Beziehung

$$\begin{aligned} &y_{s_1} = R(\{s_1\}), \quad y_{s_2} = R(\{s_1, s_2\}) - R(\{s_1\}), \ldots, \\ &y_{s_k} = R(\{s_1, \ldots, s_k\}) - R(\{s_1, \ldots, s_{k-1}\}) \end{aligned} \tag{13'}$$

und

$$y_{s_p} = 0, \quad p = (k+1), \ldots, m .$$

Nach Folgerung 3, daß $H(Q, b)$ darstellbar ist als Durchschnitt von $H''(Q,b)$ mit der Hyperebene (7), erhalten wir für $H(Q,b)$ die Ecken in eben dieser Weise mit der zusätzlichen Forderung, daß (7) erfüllt ist.

**Satz 7:** *Eine Ecke von $H(Q, b)$ ergibt sich aus einer Permutation $(i_1, \ldots, i_m)$ über die Prozedur* (13′) *mit $k = m$.*

Beweis: Jede Permutation $(i_1, \ldots, i_m)$ erzeugt nach (13′) eine Ecke von $H''(Q, b)$, außerdem gilt

$$\sum_{r=1}^{m} y_{i_r} = R(\{i_1, \ldots, i_m\}) = \sum_{j \in u(i_1, \ldots, i_m)} b_j = \sum_{j=1}^{n} b_j ,$$

da $u(\{i_1, \ldots, i_m\}) = \{1, \ldots, m\}$. ▮

**Bemerkung 1:** Auf Grund der Voraussetzung $\sum_{i=1}^{m} q_{ij} \geqq 2$, $j = 1, \ldots, n$, ergibt schon jede Menge $u(\{i_1, \ldots, i_{m-1}\}) = \{1, \ldots, m\}$.

**Bemerkung 2:** Auf Grund von Eigenschaft $\alpha$ Bedingung 2 der Funktion $R(I)$ besitzen alle Ecken des Polyeders $H(Q, b)$ nichtnegative Komponenten.

**Bemerkung 3:** Die Prozedur der Eckenkonstruktion (13) für $H(Q, b)$ nach Satz 7 entspricht der Prozedur $\varphi$ [2].

**Satz 8:** *Es existieren keine Ecken $y_0$ von $H''(Q, b)$, die mit einer Permutation $(s_1^0, \ldots, s_k^0, s_{k+1}^0, \ldots, s_m^0)$ und $k < m-1$ über* (13) *erzeugt sind, und für die $\sum_{=1}^{m} y_i^0 = \sum_{i=1}^{m} b_j$ gilt.*

Beweis: Nehmen wir an, $\mathring{y}$ sei in vorausgesetzter Form konstruiert, aber es gelte $\sum_{i=1}^{m} \mathring{y}_i = \sum_{j=1}^{n} b_j$.

Da $\mathring{y}_{s^0_{k+1}} = \ldots \mathring{y}_{s^0_m} = 0$, muß $\sum_{r=1}^{k} \mathring{y}_{s^0_r} = \sum_{j=1}^{n} b_j$ gelten. Somit wäre $u(\{s^0_1, \ldots, s^0_k\}) = \{1, \ldots, n\}$, was natürlich auch gilt für jede Menge $u(\{s^0_1, \ldots, s^0_k, s^0_{k+1}, \ldots, s^0_{k+p}\}) = \{1, \ldots, n\}, p = 1, \ldots, m-k$. Somit wird $\mathring{y}$ auch von der Permutation $(s^0_1, \ldots, s^0_m)$ mit $k = m$ über (13') bestimmt, im Widerspruch zur Voraussetzung. Die Anzahl der Permutationen $(s^0_1, \ldots, s^0_m)$ ist $m!$. ∎

**Folgerung 9:** Die Eckenanzahl $f_0(H(Q, b))$ des Polyeders $H(Q, b)$ ist höchstens $m!$.

**Bemerkung:** In [2] wird die Aussage formuliert, daß diese maximale Eckenanzahl erreicht wird, wenn die Matrix $Q$ singulär ist. Uns interessieren Verbesserungen für den Schrankenwert der Eckenzahlen. Dazu sind in [2] erste Resultate genannt, wir fügen hier eine weitere Abschätzung hinzu.

**Satz 9:** *Gilt für die Qualifikationsmatrix $Q$, daß für alle Spalten $j$ $\sum_{i=1}^{m} q_{ij} \geqq r+1$ ist, dann ist die Eckenanzahl $f_0(H(Q, b))$ durch $m!/r!$ beschränkt.*

Beweis: Aus Satz 7 und 8 folgt, daß jede der möglichen $m!$ Ecken von $H(Q, b)$ durch eine Permutation $(i_1, \ldots, i_m)$ über (13') bestimmt werden kann. Aus der Voraussetzung, daß $\sum_{i=1}^{m} q_{ij} \geqq r+1$ für alle $j$, folgt für jede Menge $\{s_1, \ldots, s_{m-r}\}$, daß $u(\{s_1, \ldots, s_{m-r}\}) = \{1, \ldots, n\}$ gilt. Setzen wir die Konstruktion einer Ecke mit einer beliebigen Menge $\{s'_{m-r+1}, \ldots, s'_m\}$ fort, so ergibt sich jeweils die gleiche Ecke $\bar{y}$, da $\bar{y}_{s'_{m-r+1}} = \ldots = \bar{y}_{s'_m} = 0$. Somit ergibt sich zu jeder Permutation $(s_1, \ldots, s_{m-r}, s'_{m-r+1} \ldots s'_m)$ bei beliebigem Rest $(s'_{m-r+1}, \ldots, s'_m)$, derer existieren $r!$, höchstens eine Ecke. Damit wird jede der $m!$ möglichen Ecken mindestens $r!$ fach erzeugt. ∎

**Definition 4:** Eine Qualifikationsmatrix $Q = \|q_{ij}\|_{m \times n}$, $n \geqq \binom{m}{r}$ heißt *r-singulär*, wenn gilt

a) für alle $j$ ist $\sum_{i=1}^{m} q_{ij} \geqq r+1$,

b) in $Q$ existiert eine Untermatrix $Q' = \|q'_{ij}\|_{m \times \binom{m}{r}}$ mit der Eigenschaft

1. $\forall j \sum_{i=1}^{m} q'_{ij} = r+1$
2. $\forall (i_1, \ldots, i_{r+1})\ \exists j_0\ q'_{i_1 j_0} = \ldots = q'_{i_{r+1} j_0} = 1$.

Im folgenden Satz werden wir zeigen, daß unter der Voraussetzung $\sum_{i=1}^{m} q_{ij} \geqq r+1$ für alle $j$, die obere Schranke $\frac{m!}{r!}$ für $r$-singuläre Matrizen erreicht wird.

**Satz 10:** *Gilt für die Qualifikationsmatrix $Q$ für alle $j$ $\sum_{i=1}^{m} q_{ij} \geqq r+1$, dann ist die Eckenanzahl $f_0(H(Q, b)) = \frac{m!}{r!}$ genau dann, wenn $Q$ eine r-singuläre Matrix ist.*

Beweis: Nach Satz 9 sind auf Grund der Voraussetzung $\sum_{i=1}^{m} q_{ij} \geqq r+1$, $j=1, \ldots, n$, höchstens $\frac{m!}{r!}$ Ecken zu erwarten. Außerdem wissen wir, daß der Teil $(t_1, \ldots, t_{m-r})$ einer Permutation $t=(t_1, \ldots, t_m)$ die Ecke $y_t$ bestimmt, da alle $y_{t_p}=0$ sind, für $p=m-r+1, \ldots, m$.

Seien $t$ und $t'$ zwei beliebige Permutationen, die sich in den ersten $m-r$ Komponenten unterscheiden. Ist $Q$ $r$-singulär, so sind dann $y(t)$ und $y(t')$ verschieden. Nehmen wir an, $t$ und $t'$ unterscheiden sich zuerst in der Komponente $t_k \neq t'_k$ (d. h. $t_i=t'_i$ $i=1, \ldots, k-1$) $(k \leqq m-r)$. Dann ist $t'_k \in \{1, \ldots, m\}$ und entspricht einem $t_s \in \{1, \ldots, m\} \setminus \{t_1, \ldots, t_k\}$. Mindestens in dieser Komponente $y_{t_s}(t)$ und $y_{t'_k}(t')$ unterscheiden sich $y(t)$ und $y(t')$.

$$y_{t_k}(t) = \sum_{j \in u(\{t_1, \ldots, t_k\})} b_j - \sum_{j \in u(\{t_1, \ldots, t_{k-1}\})} b_j$$

$$y_{t_s}(t) = \sum_{j \in u(\{t_1, \ldots, t_k, \ldots, t_s\})} b_j - \sum_{j \in u(\{t_1, \ldots, t_{s-1}\})} b_j$$

$$y_{t'_k}(t') = \sum_{j \in u(\{t'_1, \ldots, t'_k\})} b_j - \sum_{j \in u(\{t'_1, \ldots, t'_{k-1}\})} b_j,$$

wobei $t_i=t'_i$, $i=1, \ldots, k-1$ gilt.

Betrachten wir das Tupel $(t_k, t'_k=t_s, s_1, \ldots, s_{r-1})$, wobei die $s_i$ aus $\{t'_{m-r+1}, \ldots, t'_m\} \setminus \{t_k\}$ gewählt sind. Da $Q$ $r$-singulär ist, existiert eine Spalte $j_0$ mit $q_{t_k j_0}=q_{t'_k j_0}= =q_{s_1 j_0}=\ldots=q_{s_{r-1} j_0}=1$. Damit gilt $j_0 \in u(\{t_1, \ldots, t_k\})$, $j_0 \in u(\{t'_1, \ldots, t'_k\})$ und außerdem $j_0 \in u(\{t_1, \ldots, t_k, \ldots, t_p\})$ $p=k+1, \ldots, s$. Folglich erhalten wir $y_{t'_k} \geqq y_{t_s}+b_{j_0}$, d. h. $y_{t'_k} \neq y_{t_s}$. Somit gilt also, für je zwei in den ersten $m-r$ Komponenten voneinander verschiedenen Permutationen führen zu verschiedenen Ecken. Damit ergeben sich $\binom{m}{m-r}$ Möglichkeiten, $m-r$ erste Komponenten zu wählen, die dann zu $(m-r)!$ Teilpermutationen führen, somit ergeben sich $\frac{m!}{(m-r)!\,r!}(m-r)! = \frac{m!}{r!}$ verschiedene Ecken.

Beweisen wir die entgegengesetzte Richtung: Nehmen wir an, es existieren $\frac{m!}{r!}$ verschiedene Ecken (also für je eine Permutation $(t_1, \ldots, t_{m-r}, t'_{m-r+1}, \ldots, t'_m)$ mit beliebigen letzten $r$ Komponenten) und die Matrix $Q$ sei nicht $r$-singulär. $\sum_{i=1}^{m} q_{ij} \geqq \geqq r+1$ gilt nach Voraussetzung des Satzes. Betrachten wir die Untermatrix $\overline{Q'}$ von $Q$ mit $\sum_{i=1}^{m} q_{ij}=r+1$, dann gilt, da $Q$ nicht singulär ist,

$$\overline{Q'} = \|q_{ij}\|_{m \times p}, \quad p < \binom{m}{r},$$

d. h., es existiert mindestens ein Tupel $(i_1, \ldots, i_{r+1})$, für welches keine Spalte $j_0$ mit $q_{i_1 j_0} = q_{i_2 j_0} = \ldots = q_{i_{r+1} j_0} = 1$ existiert. Dann gilt für $\{i_{r+2}, \ldots, i_m\} = \{1, \ldots, m\} \setminus \{i_1, \ldots, i_{r+1}\}$, daß für alle $j$ $\sum_{p=r+2}^{m} q_{i_p j} \geqq 1$, folglich $u(\{i_{r+2}, \ldots, i_m\}) = \{1, \ldots, m\}$. Damit erhalten wir aus den Permutationen $t$ mit $t_p = i_{r+p}$, $p = 2, \ldots, m-r$, $t_{m-r} \in \{i_1, \ldots, i_{r+1}\}$ und $t_{m-r+1} \in \{i_1, \ldots, i_{r+1}\} \setminus \{t_{m-r}\}$ die gleichen Ecken $y_t$, also fallen $r+1$ in eine Ecke zusammen. Damit ist ein Widerspruch zur Existenz von $\frac{m!}{r!}$ Ecken hergeleitet. ▮

**Folgerung 10:** Für $r = 1$ ergibt sich der in [2] definierte Begriff der Singularität und der Satz, daß für 1-singuläre Matrizen $Q$ genau $m!$ Ecken in $H(Q, b)$ existieren.

Für Probleme der ganzzahligen Optimierung über Ecken von Polyedern tritt die Frage der Ganzzahligkeit des Polyeders auf. Für Polymatroide $P(R)$ folgt die Ganzzahligkeit bei der Voraussetzung, daß die Funktion $R(I)$ für alle $I \subset \{1, \ldots, m\}$ ganze Werte annimmt.

**Folgerung 11:** $H(Q, b)$ ist ein ganzzahliges Polyeder.

Für die Konstruktion lokaloptimaler, stochastischer Lösungsverfahren interessiert für die Einführung einer Metrik in der Menge aller Ecken des Polyeders $H(Q, b)$ der Begriff der *Nachbarschaft*. Entsprechend den üblichen Definitionen sind Ecken eines Polyeders benachbart, wenn sie sich in der Erfüllung einer Gleichung unterscheiden bei gleichzeitiger Erfüllung aller anderen Ungleichungen.

**Satz 11:** *Zwei Ecken $y$, $\bar{y}$ des Polyeders $H(Q, b)$ sind benachbart, wenn sich die entsprechenden Permutationen in genau einer Inversion (also $t_p = t'_{p+1}$, $t_{p+1} = t'_p$) unterscheiden.*

Beweis: Die Komponenten $y_{t_i} = \bar{y}_{t'_i}$, $i = 1, \ldots, p-1, p+1, \ldots, m$, stimmen überein, da

$$\bar{y}_{t'_i} = y_{t_i} = \sum_{j \in u(\{t_1, \ldots, t_i\})} b_j - \sum_{j \in u(\{t_1, \ldots, t_{i-1}\})} b_j, \quad i = 1, \ldots, p-1, p+1, \ldots, m.$$

Die Komponenten $y_{t_p}(y'_{t'_{p+1}})$ und $y_{t_{p+1}}(y'_{t'_p})$ unterscheiden sich

$$y_{t_p} = \sum_{j \in u(\{t_1, \ldots, t_p\})} b_j - \sum_{j \in u(\{t_1, \ldots, t_{p-1}\})} b_j$$

$$\bar{y}_{t'_p} = \sum_{j \in u(\{t_1, \ldots, t_{p-1}, t_{p+1}\})} b_j - \sum_{j \in u(\{t_1, \ldots, t_{p-1}\})} b_j$$

für $y_{t_{p+1}}$ analog.

Summiert man die Komponenten auf, so ergibt sich

$$\sum_{i=1}^{r} \bar{y}_{t'_i} = \sum_{i=1}^{r} y_{t_i} = R(\{t_1, \ldots, t_r\}), \quad r = 1, \ldots, p-1, p+1, \ldots, m.$$

Folglich erfüllen $y$ und $\bar{y}$ $m-1$ Gleichungen gemeinsam bei gleichzeitiger Erfüllung aller anderen Ungleichungen. ▮

**Bemerkung:** Bei $r$-singulären Matrizen $Q$ $(r=1, \ldots, m)$ sind auf diese Weise jeder Ecke $y(t)$ $(m-1)$ Ecken benachbart.

**Folgerung 12:** Ist $Q$ $r$-singulär, so ist $H(Q, b)$ nicht ausgeartet.

Die nach Satz 6 folgenden Aussagen lassen sich ebenfalls über das unbeschränkte Polymatroid $H'(Q, b)$ erhalten.

## 4. Facetten des Polyeders $H(Q, b)$

Wir wollen im weiteren das System von Ungleichungen (7), (9) untersuchen.

**Lemma 2:** *Sind $I^1$ und $I^2$ eine Zerlegung der Menge $I$ $(I^1 \cup I^2 = I, I^1 \cap I^2 = \emptyset, I^1, I^2 \neq \emptyset)$ mit der Eigenschaft $u(I^1) \cup u(I^2) = u(I)$, aber $u(I^1) \cap u(I^2) = \emptyset$, dann kann die Ungleichung $\sum_{i \in I} y_i \leqq R(I) = \sum_{j \in u(I)} b_j$ aus dem System (9') gestrichen werden, ohne $H''(Q, b)$ zu verändern.*

Der Beweis ergibt sich daraus, daß $\sum_{i \in I} y_i \leqq R(I)$ stets erfüllt ist, wenn folgende Ungleichungen gelten

$$\sum_{i \in I^1} y_i \leqq R(I^1) \quad \text{und} \quad \sum_{i \in I^2} y_i \leqq R(I^2) .$$

**Lemma 3:** *Ist $I' = \{i_1, \ldots, i_t\}$ eine Teilmenge von $\{1, \ldots, m\}$ mit der Eigenschaft, daß $R(I') = R(I)$ gilt, für ein $I$, $I = \{i_1, \ldots, i_t, i_{t+1}, \ldots, i_{t+s}\}$, $s = 1, \ldots, m-t$, so kann die Ungleichung $\sum_{i \in I'} y_i \leqq R(I')$ aus (9') entfernt werden.*

Beweis: Dann ist die Ungleichung $\sum_{i \in I'} y_i \leqq R(I')$ ohnehin stets erfüllt, wenn die Ungleichung

$$\sum_{i \in I} y_i \leqq R(I) \quad \text{gilt, da alle} \quad y_i \geqq 0 \quad \text{sind.} \; ▮$$

**Satz 12:** [4] *Im System (9') sind genau die Ungleichungen $\sum_{i \in I} y_i \leqq R(I)$ überflüssig, für die*

a) *eine Obermenge $I'$, $I' \supset I$, existiert mit $R(I) = R(I')$,*
b) *eine Zerlegung $I^1$, $I^2$ existiert, $I^1$, $I^2 \neq \emptyset$, $I^1 \cup I^2 = I$, $I^1 \cap I^2 = \emptyset$, mit $R(I^1) + R(I^2) = R(I)$.*

**Satz 13:** *Das Polyeder $H''(Q, b)$ besitzt $2^m - 1$ Seiten der Dimension $m-1$ genau dann, wenn $Q$ 1-singulär ist.*

Beweis: Das Polyeder $H''(Q, b)$ ist durch $2^m + m - 1$ Ungleichungen gegeben. Um die Aussagen von Satz 12 zu verwenden, zeigen wir

a) für jede Zerlegung einer Menge $I$ in Teilmengen $I^1$, $I^2 \neq \emptyset$ mit $I^1 \cap I^2 = \emptyset$, $I^1 \cup I^2 = I$ gilt

$$R(I^1) + R(I^2) > R(I) . \tag{14}$$

Betrachten wir $i_1 \in I^1$, $i_2 \in I^2$ Zeilen von $Q$, dann gilt mit der 1-Singularität der Matrix $Q$: Es existiert in $Q$ eine Spalte $j_0$ mit $q_{i_1 j_0} = q_{i_2 j_0} = 1$. Somit ist $j_0 \in u(I^1)$ und $j_0 \in u(I^2)$. Somit gilt

$$R(I^1) + R(I^2) \geqq R(I) + b_{j_0},$$

und da alle $b_j > 0$ sind, folgt die Aussage (14).

b) Für alle Mengen $I$, $|I| \leqq m-2$, gilt $R(I) < R(I')$, $I \subset I'$. Sei eine Menge $I$ gegeben, so enthalten alle Mengen $I' \supset I$ mindestens ein $i_0$ mehr (d. h. $I' \supseteqq I \cup \{i_0\}$). Da $|I| \leqq m-2$, existiert außer $i_0$ ein $i_0'$ mit $i_0' \notin I$. Auf Grund der Singularität der Matrix $Q$ existiert eine Spalte $j_0$ mit $q_{i_0 j_0} = q_{i_0' j_0} = 1$, somit ist $j_0 \in u(I')$, aber $j_0 \notin u(I)$. Folglich gilt, da alle $b_j > 0$, $R(I) < R(I')$. Betrachten wir nun die Mengen $I$, $|I| = m-1$.

Es gilt, da $\sum_{i=1}^{m} q_{ij} \geqq 2$ für alle $j$, $R(I) = R(\{1, \ldots, m\}) = \sum_{j=1}^{n} b_j$. Damit können diese Ungleichungen (genau $m$) aus (9′) entfernt werden, (oder die Ungleichungen $y_i \geqq 0$, $i = 1, \ldots, m$). Damit beschreiben $2^m - 1$ Ungleichungen das Polyeder $H''(Q, b)$. Sei $H''(Q, b)$ durch $2^m - 1$ Ungleichungen gegeben, aber $Q$ nicht 1-singulär, dann existiert eine Menge $I$, $|I| \leqq m-2$ mit

$$R(I) = R(\{1, \ldots, m\}) = \sum_{j=1}^{n} b_j ,$$

was zur Folge hat, daß eine weitere Ungleichung aus (9′) entfernt werden könnte, was einen Widerspruch bedeutet. Damit ist der Satz bewiesen. ∎

**Folgerung 13:** Das Polyeder $H(Q, b)$ besitzt $2^m - 2$ Facetten (Seiten maximaler Dimensionen, also $m-2$ dimensionale), wenn $Q$ 1-singulär ist.

**Folgerung 14:** Das Polyeder $H(Q, b)$ besitzt genau dann die maximale Eckenanzahl $f_0(H(Q, b)) = m!$, wenn es die maximale Facettenanzahl $f_{m-2}(H(Q, b)) = 2^m - 2$ besitzt.

## 5. Nichtredundante Darstellung spezieller Polyeder $H(Q, b)$

Auf der Grundlage von Satz 12 können wir nun für spezielle Matrizen $Q$ nichtredundante Systeme (9′) bzw. (7), (9) angeben. Sei $Q^1 = \|q_{ij}^1\|_{m \times n}$ mit $q_{ij}^1 = 1$, $i = 1, \ldots, m$, $j = 1, \ldots, n$.

**Satz 14:** *Das System* (9′) *von Ungleichungen, die das Polyeder* $H''(Q^1, b)$ *beschreiben, ist reduzierbar auf*

$$y_i \geqq 0, \quad i = 1, \ldots, m \tag{6}$$

$$\sum_{i=1}^{m} y_i \leqq \sum_{j=1}^{n} b_j . \tag{9''}$$

Beweis: Die Beziehung (6) ergibt sich aus $y \in E^{m+}$. Für alle Mengen $I = \{i_1\}$ erhalten wir $u(\{i_1\}) = \{1, \ldots, n\}$. Somit gilt für jede Menge $I' \supset I$, daß $R(I) = R(I') = R(\{1, \ldots, m\}) = \sum_{j=1}^{n} b_j$. Folglich sind alle diese Ungleichungen aus (9') überflüssig. Damit ergibt sich (6), (9''). ▮

**Folgerung 15:** Das Polyeder $H(Q^1, b)$ ist gegeben durch (6), (7).

**Bemerkung:** Da alle $b_j > 0$, ist $H(Q^1, b)$ ein Simplex, ein $(m-1)$-dimensionales Polyeder mit $m$ Ecken und $m$ Facetten.

Sei $Q_1$ die Qualifikationsmatrix des einparametrischen POS, $Q_1 = \|q_{ij}^{(1)}\|_{m \times m}$ hat Dreiecksstruktur:

$$q_{ij}^{(1)} = \begin{cases} 1, & i \leqq j, \quad i = 1, \ldots, m. \\ 0, & i > j, \quad j = 1, \ldots, m, \end{cases}$$

**Bemerkung:** Für $j = m$ wird die Voraussetzung $\sum_{i=1}^{m} q_{im} \geqq 2$ nicht erfüllt.

**Satz 15:** *Das nichtredundante System* (9') *für das Polyeder* $H''(Q_1, b)$ *lautet*

$$y_i \geqq 0, \quad i = 1, \ldots, m,$$

$$\sum_{i=1}^{t} y_i \leqq \sum_{j=1}^{t} b_j, \quad t = 1, \ldots, m.$$

Beweis: Die Ungleichungen $y_i \geqq 0$ folgen aus $y \in E^{m+}$. Betrachten wir nun eine beliebige Menge $I$ und $t$ sei der größte Zeilenindex der Menge $I$, dann gilt für jede Menge $I' \subset I$, $t \in I'$, daß $R(I') = R(I)$ ist. Somit sind für alle Mengen $I'$ die Ungleichungen erfüllt, wenn $\sum_{i \in I} y_i = R(I)$ erfüllt ist. Damit erhalten wir für alle Mengen $I$, mit $|I| = t$, $t$ größter Zeilenindex von $I$, genau $m$ Ungleichungen. ▮

**Folgerung 16:** Für das Polyeder $H(Q_1, b)$ ergibt sich das Ungleichungssystem

$$y_i \geqq 0, \quad i = 1, \ldots, m-1$$

$$\sum_{i=1}^{t} y_i \leqq \sum_{j=1}^{t} b_j, \quad j = 1, \ldots, m-1$$

$$y_m \geqq 0, \quad \sum_{i=1}^{m} y_i = \sum_{j=1}^{n} b_j.$$

Auf der Grundlage des Satzes 12 lassen sich in dieser Weise für beliebige Matrizen $Q$ die nichtredundanten Systeme (9') erarbeiten.

### Literatur


[1] Kowaljow, M. M.; E. Girlich: Zum Problem der optimalen Standardisierung. MOS, Series Optimization vol. 8 (1977) 1, 89–103.

[2] Гирлих, Э., и М. М. Ковалев: Класс многогранников задачи стандартизации с максимальным числом вершин. Изв. АН БССР, сер. физ.-мат. наук, (1974) № 6.

[3] GIRLICH, E.; M. M. KOWALJOW; W. A. PIRJANOWITSCH: Lokaloptimale stochastische Lösungsverfahren. Thesen zur Tagung „Mathematische Methoden in der Ökonomie", Vitte, Okt. 1977.
[4] EDMONDS, J.: Submodular functions, matroids and certain polyhedra. Combinatorical Structures and their applications, Calgary Intern. Conf., Canada, 1969.
[5] BARTELS, A.: A priori-Informationen zur linearen Optimierung. Verlag A. Heim, Meisenheim 1973.
[6] КОВАЛЕВ, М. М.: Дискретная оптимизация. Изд. БГУ, Минск 1977.
[7] ЕМЕЛИЧЕВ, В. А., М. М. КОВАЛЕВ и М. А. КРАВЦОВ: Многогранники, графы, оптимизация, « Наука » 1980.
[8] GRÖTSCHEL, M.; M. PADBERG: Lineare Charakterisierungen von Traveling Salesman Problemen. Z. f. Oper. Res. **21** (1977) No 1.
[9] ФОРД, Л., и Д. ФАЛКЕРСОН: Потоки в сетях. « Наука » 1966.
[10] NOŽIČKA, F.; J. GUDDAT; H. HOLLATZ: Einführung in die Theorie der linearen Optimierung. Akademieverlag Berlin 1972.
[11] GIRLICH, E.; M. M. KOWALJOW: Minimierung konvexer Funktionen über Polymatroiden. Tagung „Optimierung – Theorie und Anwendung", Eisenach 1978.

**Резюме**

В этой работе исследуются задачи оптимальной стандартизации. Предложенные результаты о структуре допустимой области решений являются основой алгоритмов решений такой класс задач.

**Summary**

In this paper we investigate problems of optimal unification. We present results about the structure of admissible solutions set. Methods for solution of these problems are founded on the given results.

Received June 1979; revised November 1979.

## Book Review

W. Prenowitz; J. Jantosciak: **Join Geometries. A Theory of Convex Sets and Linear Geometry.** Springer-Verlag, Berlin-Heidelberg-New York 1979, 532 S., DM 42,–.

Das vorliegende Buch ist der erste „undergraduate Text in Mathematics“, in welchem dem Leser die klassische lineare Geometrie durch den Zugang über die Verbindungsoperation (Join Operation), die je zwei Punkten ein sogenanntes Segment zuordnet, erschlossen wird. Die klassische Euklidische Geometrie erweist sich dabei als eine Theorie der konvexen Mengen und der konvexen Mengenfamilien. Das der Theorie zu Grunde gelegte Axiomensystem besteht aus Axiomen, die elementare Eigenschaften der Verbindungsoperation enthalten. Diese sind formuliert als universelle Eigenschaften von Punkten. Hervorgehoben werden muß, daß die benutzten Axiome koordinaten- und dimensionsfrei sind, womit die aus den Axiomen hergeleiteten Aussagen anwendbar sind auf Geometrien beliebiger Dimension, sowohl unendlicher als auch endlicher.

Nachdem im Kapitel 1 die Eigenschaft der Verbindungsoperation und der aus der Verbindungsoperation abgeleiteten Ausdehnungsoperation (Extension Operation) am Modell der Euklidischen Geometrie ausführlich beschrieben werden, wird in den Kapiteln 2 bis 6 die abstrakte Theorie auf der Grundlage des Begriffes der Verbindungsoperation entwickelt. In Kapitel 5 wird dabei der Begriff der Verbindungsgeometrie (Join Geometry) eingeführt. Eine Verbindungsgeometrie ist ein Modell der Theorie; es ist ein Paar $(J, \cdot)$, erzeugt durch eine beliebige nichtleere Menge $J$ und einer Verbindungsoperation $\cdot$ in $J$, das einer Grundmenge von Axiomen genügt. Es wird eine umfangreiche Liste von Modellen für Verbindungsgeometrien angegeben und der Begriff des Isomorphismus von Verbindungsgeometrien behandelt. Der Untersuchung spezieller Mengen in Verbindungsgeometrien sind die Kapitel 6 bis 9 gewidmet. In diesen werden lineare Mengen, Extremalmengen, strahlenförmige Mengen sowie Kegel und Halbräume untersucht. In den Kapiteln 10 bis 13 behandeln die Autoren spezielle Verbindungsgeometrien, so in Kapitel 10 Faktorgeometrien (Factor Geometry), in Kapitel 11 Austauschgeometrien (Exchange Geometry) und in den Kapiteln 12 und 13 geordnete Verbindungsgeometrien (Ordered Join Geometry). Die Kapitel 7 bis 13 können zum Teil unabhängig voneinander studiert werden. Dabei nehmen die Autoren jedoch in Kauf, daß Begriffe mehrfach definiert werden; man vgl. etwa die Seiten 305 und 351. Eine Strukturtabelle, die die Zusammenhänge der einzelnen Kapitel veranschaulicht, ist im Buch enthalten. Positiv vermerkt werden muß, daß der gesamte Text begleitet wird durch eine umfangreiche Kollektion von Aufgaben und Problemstellungen. Bei einer sorgfältigen Durchsicht fällt auf, daß die in der Einleitung mehrfach erwähnten Zusammenhänge zu algebraischen Strukturen im Buch nicht explizit behandelt werden.

Das Studium des Buches setzt einige formale Kenntnisse in der Geometrie sowie Grundkenntnisse auf dem Gebiet der Algebra und der Mengentheorie voraus. Es ist für Studenten und Mathematiker gleichermaßen interessant. Dem Spezialisten auf dem Gebiet der Geometrie sollte der moderne Zugang zur klassischen Geometrie interessieren. Für Spezialisten auf dem Gebiet der Theorie der konvexen Mengen könnte der weite Einblick, den diese Theorie in die Theorie der konvexen Mengen vermittelt, von Interesse sein. Der Inhalt von Kapitel 1 läßt sich durchaus in Arbeitsgemeinschaften „Junger Mathematiker“ behandeln.

G. Heidekrüger

Math. Operationsforsch. Statist., Ser. Optimization, Vol. 11 (1980) No. 4, 563–578

# On the Quasi-solution of Overdetermined Ill-conditioned Systems of Linear Equations and its Realization by means of Parametric Linear Programming

R. TICHATSCHKE[1] and B. HOFMANN[1]

**Summary:** The paper is concerned with the stability of the $L_\infty$-quasi-solutions in the sense of IVANOV for overdetermined, ill-conditioned linear equation systems. A stability theorem is proved, which considers perturbations in the coefficient-matrix of the equation system. Then an efficient way for the computation of $L_\infty$-quasi-solutions is shown by means of parametric linear programming. The numerical investigations performed allow statements on the stability behaviour of such solutions when the right-hand side and the matrix of the equation system are perturbed. It is shown that $L_\infty$-quasi-solutions just provide good approximations if EUCLIDean-quasi-solutions or equivalent regularized solutions are very imprecisely.

## 1. Introduction

The mathematical treatment of problems for the interpretation of measurement data in order to reconstruct physical quantities often requires the solution of FREDHOLM integral equations of the first kind

$$\int_a^b K(s, t)\, f(t)\, \mathrm{d}t = g(s)\ . \tag{1.1}$$

Considering the solution function $f(t)$ and the function of measurable values $g(s)$ in the space of continuous or quadratic integrable functions defined on an interval, the problem (1.1) becomes incorrect or improperly posed in the sense of HADAMARD's definition (see TYKHONOV/ARSENIN [14]). There do not exist any solutions $f(t)$ of equation (1.1) for every function $g(s)$ with $s \in [c, d]$. Moreover, small perturbations of the right-hand term $g(s)$, which arise from measurement errors, may cause arbitrarily great deviations from the exact solution.

If we discretisize the integral equation (1.1) by an overdetermined system of linear equations, we find

$$\boldsymbol{A}\boldsymbol{x} = \boldsymbol{b} \quad \boldsymbol{x} \in R^n, \quad \boldsymbol{b} \in R^m, \quad m \geqq n, \quad \operatorname{rank}(\boldsymbol{A}) = n \tag{1.2}$$

as approximation of the integral equation problem. The algebraic system (1.2) possesses an ill-conditioned matrix $\boldsymbol{A}$, which represents a finite-dimensional approximation of the integral operator in the incorrect problem (1.1). FRIEDRICH et al.

[1] Sektion Mathematik der TH Karl-Marx-Stadt, DDR - 9010 Karl-Marx-Stadt, Postfach 964.

give a survey of the possibilities of using different kinds of additional information for the computation of numerically stable solutions of such ill-conditioned systems (1.2) (see [7]).

The method of quasi-solution introduced by IVANOV in [9] and [10] consists in the restriction of possible solutions of an incorrect or ill-posed problem to a compact set (comp. CHAVENT [4]). Therefore approximate solutions so obtained stably depend on the input data. Applying this method to (1.2), we have to find the set $\{\boldsymbol{x}(\nu)\}$ of vectors $\boldsymbol{x} \in R^n$ from the compact $C = \{\boldsymbol{x} \in R^n \mid \|\boldsymbol{x}\| \leqq \nu\}$, which minimize the defect

$$\{\boldsymbol{x}(\nu)\} = \{\boldsymbol{x} \in C \mid \|\boldsymbol{A}\boldsymbol{x} - \boldsymbol{b}\| = \min!\}\,. \tag{1.3}$$

Using the projection $\mathsf{P}_A \boldsymbol{b} = \{\boldsymbol{y} = \boldsymbol{A}\boldsymbol{x},\ \boldsymbol{x} \in C \mid \|\boldsymbol{y} - \boldsymbol{b}\| = \min!\}$ and the MOORE-PENROSE-pseudoinverse $\boldsymbol{A}^+ = (\boldsymbol{A}^\mathsf{T}\boldsymbol{A})^{-1}\boldsymbol{A}^\mathsf{T}$ of the full-rank matrix $\boldsymbol{A}$, the set of quasi-solutions is characterized by the formula $\{\boldsymbol{x}(\nu)\} = \{\boldsymbol{A}^+ \mathsf{P}_A \boldsymbol{b}\}$. If the vector-norms in $R^m$ and $R^n$ are chosen in such a way that the compact set of images $\boldsymbol{A}C = \{\boldsymbol{y} \in R^m \mid \boldsymbol{y} = \boldsymbol{A}\boldsymbol{x},\ \boldsymbol{x} \in C\}$ forms a strictly convex set, IVANOV [9] proves the continuous dependence of the uniquely determined quasi-solution $\boldsymbol{x}(\nu)$ on the right-hand side $\boldsymbol{b}$.

In the case of EUKLIDean vector-norms $\|\cdot\|_2$, statements on the behaviour of the quasi-solution are easy to obtain. There the formula

$$\boldsymbol{x}(\nu) = \begin{cases} (\boldsymbol{A}^\mathsf{T}\boldsymbol{A} + \alpha \cdot \boldsymbol{I})^{-1}\boldsymbol{A}^\mathsf{T}\boldsymbol{b} \ \text{ with } \alpha(\nu) : \|\boldsymbol{x}(\nu)\|_2 = \nu & \text{if } \nu < \|\boldsymbol{A}^+\boldsymbol{b}\|_2 \\ \boldsymbol{A}^+\boldsymbol{b} & \text{if } \nu \geqq \|\boldsymbol{A}^+\boldsymbol{b}\|_2 \end{cases} \tag{1.4}$$

holds. For large compact sets $C$, the quasi-solution is equivalent to the pseudonormal-solution $\boldsymbol{A}^+\boldsymbol{b}$. On the other hand, for smaller values $\nu$, the method of quasi-solution represents a special case of TYKHONOV's regularization method (comp. [14]). Then the quasi-solution $\boldsymbol{x}(\nu)$ minimizes the functional

$$F(\boldsymbol{x}) = \|\boldsymbol{A}\boldsymbol{x} - \boldsymbol{b}\|_2^2 + \alpha(\nu)\,\|\boldsymbol{x}\|_2^2 \quad \text{with} \quad \boldsymbol{x}(\nu) = \boldsymbol{x}(\nu, \alpha(\nu)) = \min_{\boldsymbol{x} \in R^n} F(\boldsymbol{x}) \tag{1.5}$$

for a regularization parameter $\alpha(\nu) > 0$ that is uniquely determined by the equation $\|\boldsymbol{x}(\nu, \alpha(\nu))\|_2 = \nu$ when $\nu < \|\boldsymbol{A}^+\boldsymbol{b}\|_2$. Using a singular-value-decomposition

$$\boldsymbol{A} = \begin{pmatrix} \boldsymbol{v}_1^\mathsf{T} \\ \vdots \\ \boldsymbol{v}_m^\mathsf{T} \end{pmatrix} \begin{pmatrix} s_1 & & 0 \\ & \ddots & \\ 0 & & s_n \\ & 0 & \end{pmatrix} (\boldsymbol{u}_1 \ldots \boldsymbol{u}_n) = \boldsymbol{V}\boldsymbol{S}\boldsymbol{U} \tag{1.6}$$

of the matrix $\boldsymbol{A}$, in which $\boldsymbol{U}$ and $\boldsymbol{V}$ denote unitary matrices and $s_1 \geqq s_2 \geqq \ldots \geqq$ $\geqq s_n > 0$ are the positive singular values of $\boldsymbol{A}$ (comp. KUHNERT [11], p. 35), we get the relation

$$\boldsymbol{x}(\nu) = \begin{cases} \displaystyle\sum_{i=1}^n \frac{1}{s_i} \cdot \frac{s_i^2}{s_i^2 + \alpha(\nu)} \cdot (\boldsymbol{b}, \boldsymbol{v}_i)\,\boldsymbol{u}_i & \text{with } \alpha(\nu) : \displaystyle\sum_{i=1}^n \frac{(\boldsymbol{b}, \boldsymbol{v}_i)^2 s_i^2}{(s_i^2 + \alpha(\nu))^2} = \nu^2 \\ \displaystyle\sum_{i=1}^n \frac{1}{s_i} \cdot (\boldsymbol{b}, \boldsymbol{v}_i)\,\boldsymbol{u}_i & \text{with } \displaystyle\sum_{i=1}^n \frac{(\boldsymbol{b}, \boldsymbol{v}_i)^2}{s_i^2} \leqq \nu\,. \end{cases} \tag{1.7}$$

If the real number $\nu$ decreases, the corresponding parameter $\alpha(\nu)$ increases. As a result the factor of damping $\frac{s_i^2}{s_i^2+\alpha(\nu)}<1$ decreases and the stability of the solution process, which is characterized by the smallest possible deviation $\|\boldsymbol{x}(\nu)-\boldsymbol{x}_\delta(\nu)\|_2$, grows. Here we suppose that the inequality $\|\boldsymbol{b}-\boldsymbol{b}_\delta\|_2<\delta$ holds with a small error bound $\delta$. On the other hand, the pseudonormal-solution $\boldsymbol{A}^+\boldsymbol{b}$ will be instable, because the condition number cond $(\boldsymbol{A})=s_1\cdot s_n^{-1}$ has a large value and thus the smallest singular value $s_n$ of the matrix $\boldsymbol{A}$ is close to zero. However, for small values $\nu$, the stable dependence on the right-hand side must often be achieved at the expense of an essential deviation of the quasi-solution from the solution $\bar{\boldsymbol{x}}$ to be reconstructed. To find a convenient compact set $C$, a compromise between stability and precision of the approximate solution is necessary.

If we deal with the problem (1.2) by means of $L_\infty$-norms, then the set $\boldsymbol{A}C$ is not generally strictly convex. Therefore the uniqueness property of the quasi-solution gets lost. Techniques of spectral theory require Hilbert-space norms and are not suitable for the investigation of the problem in this case. Ivanov [10] proves the existence of such a real number $\delta(\varepsilon, \boldsymbol{A}, \nu)$ for any small values $\varepsilon>0$ and for an arbitrary element of the set of quasi-solutions $\{\boldsymbol{x}_\delta(\nu)\}$ that at least one quasi-solution $\boldsymbol{x}(\nu)$ satisfies the inequality $\|\boldsymbol{x}(\nu)-\boldsymbol{x}_\delta(\nu)\|_\infty<\varepsilon$ for a perturbed right-hand side $\boldsymbol{b}_\delta$ with $\|\boldsymbol{b}-\boldsymbol{b}_\delta\|_\infty<\delta$. The next section contains the proof of a stability theorem for the method of quasi-solution, in which perturbations of the matrix $\boldsymbol{A}$ are additionally considered. Furthermore, by means of parametric linear programming a way of calculating quasi-solutions in the sense of $L_\infty$-norms is developed. Numerical experiments allow statements on the stability behaviour of such solutions with perturbed right-hand vector $\boldsymbol{b}$ and system matrix $\boldsymbol{A}$ and on optimal choice of the parameter $\nu$. By means of analytical investigations such statements cannot be made. Experiments show that the quasi-solution is stable, just when the parameter $\nu=\underline{\nu}$ is chosen such that $\boldsymbol{x}(\underline{\nu})$ represents a Chebyshev-solution of $\{\boldsymbol{A}\boldsymbol{x}=\boldsymbol{b}, \boldsymbol{x}\in C\}$.

## 2. A Stability Theorem

Besides perturbations of the right-hand side by reason of measurement errors, small perturbations of the system matrix $\boldsymbol{A}$ are unavoidable in the numerical realization of quasi-solutions. Because of inaccurate knowledge of $\boldsymbol{A}$ and the rounding of input data in the computer instead of the quasi-solution $\boldsymbol{x}_\delta(\nu)$ with $\boldsymbol{b}=\boldsymbol{A}\bar{\boldsymbol{x}}$, $\|\boldsymbol{b}-\boldsymbol{b}_\delta\|<\delta$ we calculate the quasi-solution

$$\boldsymbol{x}_\delta^h(\nu)=\{\boldsymbol{x}\in C \mid \|\boldsymbol{A}^h\boldsymbol{x}-\boldsymbol{b}_\delta\|=\min!\}$$

with the perturbed matrix $\boldsymbol{A}^h$ and $\|\boldsymbol{A}^h-\boldsymbol{A}\|\leqq h$. There $\|\boldsymbol{x}\|$ denotes an arbitrary vector-norm and $\|\boldsymbol{A}\|$ a matrix-norm, which is consistent with this vector-norm. The symbol $\|\boldsymbol{A}\|_2$ marks the spectral-norm. Furthermore, we assume the inequality $\|\bar{\boldsymbol{x}}\|\leqq\nu$. On condition of the above assumptions the following stability theorem holds:

**Theorem:** *For any small value $\varepsilon>0$, there are two positive numbers $\delta$ and $h$ which* *may depend on ε, $\boldsymbol{A}$ and ν such that the inequality*

$$\|\boldsymbol{x}_\delta^h(\nu)-\bar{\boldsymbol{x}}\|<\varepsilon \quad \textit{if} \quad \|\boldsymbol{b}-\boldsymbol{b}_\delta\|<\delta \quad \text{and} \quad \|\boldsymbol{A}-\boldsymbol{A}^h\|\leqq h$$

*is satisfied for any quasi-solution* $\boldsymbol{x}_\delta^h(\nu)$.

Proof: We prove the theorem by means of the concept of quasi-distance $\beta(\mathfrak{A},\mathfrak{B})=\sup_{\boldsymbol{a}\in\mathfrak{A}}\inf_{\boldsymbol{b}\in\mathfrak{B}}\|\boldsymbol{a}-\boldsymbol{b}\|$ between two sets $\mathfrak{A}$ and $\mathfrak{B}$ of a metric space. We show the β-continuity $\beta(\{\boldsymbol{x}_\delta^h(\nu)\},\{\boldsymbol{x}(\nu)\})<\varepsilon$ for $\|\boldsymbol{b}-\boldsymbol{b}_\delta\|<\delta$ and $\|\boldsymbol{A}-\boldsymbol{A}^h\|\leqq h$ of the set of quasi-solutions $\{\boldsymbol{x}_\delta^h(\nu)\}$ relative to the solution vector $\{\boldsymbol{x}(\nu)\}=\{\bar{\boldsymbol{x}}\}$. Therefore this proves the statement. We apply the inequality

$$\beta(\mathfrak{A},\mathfrak{B})\leqq\beta(\mathfrak{A},\mathfrak{C})+\beta(\mathfrak{C},\mathfrak{B}), \tag{2.1}$$

which results for three sets $\mathfrak{A}$, $\mathfrak{B}$ and $\mathfrak{C}$ of a metric space from the inequality

$$\inf_{\boldsymbol{b}\in\mathfrak{B}}\|\boldsymbol{a}-\boldsymbol{b}\|\leqq\|\boldsymbol{a}-\boldsymbol{c}\|+\inf_{\boldsymbol{b}'\in\mathfrak{B}}\|\boldsymbol{c}-\boldsymbol{b}'\| \quad \text{with} \quad \boldsymbol{a}\in\mathfrak{A} \quad \text{and} \quad \boldsymbol{c}\in\mathfrak{C}$$

(comp. [5], p. 44). Thus we get

$$\beta(\{\boldsymbol{x}_\delta^h(\nu)\},\{\bar{\boldsymbol{x}}\})\leqq\beta(\{\boldsymbol{x}_\delta^h(\nu)\},\{\boldsymbol{x}_\delta(\nu)\})+\beta(\{\boldsymbol{x}_\delta(\nu)\},\{\bar{\boldsymbol{x}}\}). \tag{2.2}$$

IVANOV [9] proves that for any $\varepsilon>0$ there is a $\delta_0(\varepsilon,\boldsymbol{A},\nu)>0$ with $\beta(\{\boldsymbol{x}_\delta(\nu)\},\{\bar{\boldsymbol{x}}\})<\frac{\varepsilon}{2}$ for $\|\boldsymbol{b}-\boldsymbol{b}_\delta\|<\delta_0$. In the following we show that

$$\beta(\{\boldsymbol{x}_\delta^h(\nu)\},\{\boldsymbol{x}_\delta(\nu)\})<\frac{\varepsilon}{2}$$

holds, if both conditions

$$\beta(\{\mathsf{P}_{A^h}\boldsymbol{b}_\delta\},\{\mathsf{P}_A\boldsymbol{b}_\delta\})<\frac{\varepsilon\cdot s_n}{4}$$

and

$$\|\boldsymbol{A}-\boldsymbol{A}^h\|\leqq\min\left(\frac{s_n^2}{4s_1},\frac{\varepsilon}{32s_1\nu\left(\frac{1}{s_n^2}+\frac{3s_1^2}{s_n^4}\right)}\right)$$

are satisfied.

With $\|\boldsymbol{A}-\boldsymbol{A}^h\|\leqq\frac{s_n^2}{4s_1}$ for the smallest singular value $s_n^h$ of the matrix $A^h$, we get $s_n^h\geqq\frac{s_n}{2}$ and thus rank $(\boldsymbol{A}^h)=n$. Obviously, $\|\boldsymbol{A}^h\|\leqq\|\boldsymbol{A}\|_2$ holds and also

$$\|\boldsymbol{A}^{h\mathsf{T}}\boldsymbol{A}^h-\boldsymbol{A}^\mathsf{T}\boldsymbol{A}\|_2\leqq3\|\boldsymbol{A}^h-\boldsymbol{A}\|\cdot\|\boldsymbol{A}\|_2\leqq\frac{3s_n^2}{4}. \tag{2.3}$$

From $s_i^2-\frac{3s_n^2}{4}\leqq(s_i^h)^2\leqq s_i^2+\frac{3s_n^2}{4}$ for $i=1(1)\,n$ we have $(s_n^h)^2\geqq\frac{s_n^2}{4}$. Therefore for any two vectors $\boldsymbol{z}^h\in\{\mathsf{P}_{A^h}\boldsymbol{b}_\delta\}$ and $\boldsymbol{z}\in\{\mathsf{P}_A\boldsymbol{b}_\delta\}$ there exist uniquely determined quasi-solutions $\boldsymbol{A}^{h+}\boldsymbol{z}^h$ and $\boldsymbol{A}^+\boldsymbol{z}$ with

$$\|\boldsymbol{A}^{h+}\boldsymbol{z}^h-\boldsymbol{A}^+\boldsymbol{z}\|\leqq\|\boldsymbol{A}^{h+}-\boldsymbol{A}^+\|\,\|\boldsymbol{A}^h\|\,\nu+\|\boldsymbol{A}^+\|\,\|\boldsymbol{z}^h-\boldsymbol{z}\|$$
$$\leqq2\,\|\boldsymbol{A}^{h+}-\boldsymbol{A}^+\|\cdot s_1\cdot\nu+\frac{1}{s_n}\|\boldsymbol{z}^h-\boldsymbol{z}\|\leqq2\,\|\boldsymbol{A}^{h+}-\boldsymbol{A}^+\|\cdot s_1\cdot\nu+\frac{\varepsilon}{4}.$$

Because of $\|A - A^h\| \leqq \varepsilon \cdot \left(32 s_1 \nu \left(\frac{1}{s_n^2} + \frac{3 s_1^2}{s_n^4}\right)\right)^{-1}$ the inequality

$$\begin{aligned}
\|A^{h+} - A^+\|_2 &= \|(A^{h\mathsf{T}} A^h)^{-1} A^{h\mathsf{T}} - (A^\mathsf{T} A)^{-1} A^\mathsf{T}\|_2 \\
&\leqq \|(A^{h\mathsf{T}} A^h)^{-1}\|_2 \cdot \|A - A^h\|_2 + \|(A^{h\mathsf{T}} A^h)^{-1} - (A^\mathsf{T} A)^{-1}\|_2 \cdot \|A\|_2 \\
&\leqq \frac{4}{s_n^2} \|A - A^h\|_2 + \frac{4}{s_n^2} \cdot \frac{1}{s_n^2} \cdot \|A^{h\mathsf{T}} A^h - A^\mathsf{T} A\|_2 \cdot s_1 \\
&\leqq \frac{4}{s_n^2} \|A - A^h\|_2 + \frac{12}{s_n^4} \cdot s_1^2 \|A - A^h\|_2 = \left(\frac{4}{s_n^2} + \frac{12 s_1^2}{s_n^4}\right) \cdot \|A - A^h\|_2 \leqq \frac{\varepsilon}{8 s_1 \nu}
\end{aligned}$$

and thus the estimation $\beta(\{x_\delta^h(\nu)\}, \{x_\delta(\nu)\}) < \frac{\varepsilon}{2}$ holds. Moreover, we consider the inequality

$$\beta(\{\mathsf{P}_{A^h} b_\delta\}, \{\mathsf{P}_A b_\delta\}) \leqq \beta(\{\mathsf{P}_{A^h} b_\delta\}, \{\mathsf{P}_A b\}) + \beta(\{\mathsf{P}_A b\}, \{\mathsf{P}_A b_\delta\}). \tag{2.4}$$

Using the theorem in [10], there is a value $\delta_1(\varepsilon, A, \nu) > 0$ such that for $\|b - b_\delta\| < \delta_1$ the estimation

$$\beta(\{\mathsf{P}_A b\}, \{\mathsf{P}_A b_\delta\}) = \inf_{z \in \{\mathsf{P}_A b_\delta\}} \|\bar{x} - z\| \leqq \sup_{z \in \{\mathsf{P}_A b_\delta\}} \|\bar{x} - z\| = \beta(\{\mathsf{P}_A b_\delta\}, \{\mathsf{P}_A b\}) < \frac{\varepsilon \cdot s_n}{8}$$

holds. We get

$$\begin{aligned}
&\beta(\{\mathsf{P}_{A^h} b_\delta\}, \{\mathsf{P}_A b\}) \leqq \beta(\{\mathsf{P}_{A^h} b_\delta\}, \{b_\delta\}) + \|b - b_\delta\| \\
&\leqq \|b_\delta - A^h \bar{x}\| + \|b - b_\delta\| \leqq \|A\bar{x} - A^h \bar{x}\| + 2\|b - b_\delta\| \leqq \|A^h - A\| \nu + 2\|b - b_\delta\|.
\end{aligned}$$

The inequality $\beta(\{\mathsf{P}_{A^h} b_\delta\}, \{\mathsf{P}_A b_\delta\}) \leqq \frac{\varepsilon \cdot s_n}{4}$ is satisfied on the condition that $\|b - b_\delta\| < \min\left(\frac{\varepsilon \cdot s_n}{32}, \delta_1\right)$ and $\|A - A^h\| \leqq \frac{\varepsilon \cdot s_n}{16 \nu}$. If the vector $b$ and the matrix $A$ are so perturbed that $\|b - b_\delta\| < \min\left(\delta_0, \delta_1, \frac{\varepsilon \cdot s_n}{32}\right)$ and

$$\|A - A^h\| \leqq \min\left(\frac{s_n^2}{4 s_1}, \frac{\varepsilon \cdot s_n}{16 \nu}, \frac{\varepsilon}{32 s_1 \nu \left(\frac{1}{s_n^2} + \frac{3 s_1^2}{s_n^4}\right)}\right)$$

and quasi-solution $x_\delta^h(\nu)$ satisfies the inequality $\|x_\delta^h(\nu) - \bar{x}\| < \varepsilon$. ∎

The above theorem ensures the quality of the method of quasi-solution in the case of very small perturbations $\delta$ and $h$. However, for many applications the precision of $b_\delta$ and $A^h$ is determined by the problem itself and cannot be improved essentially. To get stable approximate solutions $x_\delta^h(\nu)$ relative to the system matrix and the right-hand side, the compact set $C$ and thus the number $\nu$ must be reduced. In that case the inequality $\|\bar{x}\| \leqq \nu$ may be perturbed. Then, for the sake of stability, we take an error $\|x_\delta^h(\nu) - \bar{x}\| \geqq \|\bar{x}\| - \nu$ of the quasi-solution for granted.

The present situation is analogous to the application of TYKHONOV's regularization method, in which for fixed error bounds $\delta$ and $h$ the regularization parameter $\alpha$ must be chosen sufficiently large. TYKHONOV and ARSENIN [14], p. 83 offer a few useful empirical criteria for choosing parameter $\alpha$. The following investigations of parametric linear programming using $L_\infty$-norms result in a criterion for a useful choice of parameter $\nu$. That criterion is also of an empirical nature, but it has proved it's usefulness in many examples.

## 3. The Problem of Parametric Linear Programming

To solve large systems of the kind (1.2) numerically, it is convenient to use $L_\infty$-norms instead of EUCLIDean norm (comp. BARTELS/GOLUB [3], ROSEN [13], ANTON/DURIS [2] et al.). We define the defect vector

$$\boldsymbol{d}=\boldsymbol{d}(\boldsymbol{x})=\boldsymbol{A}\boldsymbol{x}-\boldsymbol{b} \quad \text{with} \quad \|\boldsymbol{d}\|_\infty=\max_i |d_i| .$$

From (1.3) the following problem is derived: find a quasi-solution $\boldsymbol{x}=\boldsymbol{x}(\nu)$ such that

$$\|\boldsymbol{d}(\boldsymbol{x}(\nu))\|_\infty=\min \{\|\boldsymbol{d}\|_\infty \mid \boldsymbol{d}=\boldsymbol{A}\boldsymbol{x}-\boldsymbol{b},\ \|\boldsymbol{x}\|_\infty \leqq \nu\} . \tag{3.1}$$

$$\tilde{\boldsymbol{A}}=\left[\begin{array}{c|c|c|c} \boldsymbol{A}^\mathsf{T} & -\boldsymbol{A}^\mathsf{T} & -\boldsymbol{E}_n & \boldsymbol{E}_{n+1} \\ 1\ \ldots\ 1 & 1\ \ldots\ 1 & 0 & \end{array}\right], \quad \begin{array}{l} \tilde{\boldsymbol{x}}=[\boldsymbol{x} \vdots x_0]^\mathsf{T} \in \boldsymbol{R}^{n+1}, \\ \tilde{\boldsymbol{c}}=[0, \ldots, 0, 1] \in \boldsymbol{R}^{n+1}, \end{array}$$

$$\tilde{\boldsymbol{b}}=(b_1, \ldots, b_m \mid -b_1, \ldots, b_m \mid -\nu, \ldots, -\nu \mid -\nu, \ldots -\nu, 0)^\mathsf{T} .$$

Problem (3.1) may now be written in the equivalent form

$$\min \{(\tilde{\boldsymbol{c}}, \tilde{\boldsymbol{x}}) : \tilde{\boldsymbol{A}}^\mathsf{T}\tilde{\boldsymbol{x}} \geqq \tilde{\boldsymbol{b}}\} \tag{3.2}$$

and the dual problem arising from this is

$$\max \{(\tilde{\boldsymbol{b}}, \boldsymbol{u}) \mid \tilde{\boldsymbol{A}}\boldsymbol{u}=\tilde{\boldsymbol{c}},\ \boldsymbol{u} \geqq 0\} . \tag{3.3}$$

We can easily see that a feasible solution (quasi-solution to (1.2)) always exists for $\nu \geqq 0$ in the problem (3.2) and the objective function is bounded below. Therefore the problem (3.2) is solvable for all $\nu \in \mathfrak{A} := \{\nu \mid 0 \leqq \nu < \infty\}$. The function

$$\varphi(\nu)=\min \{(\tilde{\boldsymbol{c}}, \tilde{\boldsymbol{x}}) \mid \tilde{\boldsymbol{x}} \in R(\nu)\} \quad \text{with} \quad R(\nu)=\{\tilde{\boldsymbol{x}} \in \boldsymbol{R}^{n+1} \mid \tilde{\boldsymbol{A}}^\mathsf{T}\tilde{\boldsymbol{x}} \geqq \tilde{\boldsymbol{b}}\} \tag{3.4}$$

is the solution-function of problem (3.2).

We now obtain a solution $\tilde{\boldsymbol{x}}(\nu)=(\boldsymbol{x}(\nu) \mid x_0(\nu))^\mathsf{T}$ of (3.2), which describes the behaviour of the defect-value $\|\boldsymbol{d}(\boldsymbol{x}(\nu))\|_\infty=x_0(\nu)$ by the solution-function, because $\varphi(\nu)= =x_0(\nu)$. The possible decrease in error by increasing the norm of the quasi-solution $\boldsymbol{x}(\nu)$ is completely described by the solution-function (3.4).

**Lemma 1:** *The solution-function $\varphi(\nu)$ in (3.4) is a piecewise linear, convex and non-increasing function over the set $\mathfrak{A}$.*

The piecewise linearity and convexity follows from the well known results of the linear parametric optimization theory (NOŽIČKA et al. [12], Chapter 3.2.). The monotonicity follows from the fact that $R(\nu_2) \supset R(\nu_1)$, if $\nu_2 > \nu_1$.

We now consider the problem (3.3). The parameter $\nu$ appears in the objective function of (3.3) and therefore a class of an infinite number of linear programming problems for $\nu \in [0, \infty)$ is defined, that is, (3.3) can be solved by algorithms for parametric linear programming problems. Essentially the computation of any optimal solution of (3.3) for a fixed parameter-value $\nu = \nu^*$ requires only that the inverse for a selected $(n+1)\times(n+1)$-matrix can be computed. This usually requires approximate $2n$ simplex pivot operations. Thus the computation time depends primarily on $n$ and only to a small extent on the value of $m$. The simplex method for the computation of quasi-solutions in (1.2) will therefore be most efficient when $n \ll m$ and when problem (3.3) is used for the computation of the problem (3.1). By the duality theory of linear programming problems we obtain with an optimal solution $\boldsymbol{u}^*$ of (3.3) for the parameter-value $\nu = \nu^*$ an optimal solution of (3.2), together with the corresponding dual-variables of this problem (denoted by $\tilde{\boldsymbol{x}}^* = (\boldsymbol{x}^*, x_0^*)^{\mathsf{T}}$). An initial basis for the start of the simplex-algorithm for (3.3) is given by the unity matrix $\boldsymbol{E}_{n+1}$ which is contained in the matrix $\tilde{\boldsymbol{A}}$. An optimal basis for $\nu = 0$ will be obtained by replacing one or more columns of $\boldsymbol{E}_{n+1}$ by columns selected from the first $(2m+n)$ columns of $\tilde{\boldsymbol{A}}$. For the numerical treatment we can use $\varepsilon$-techniques in order to overcome the extreme degeneration of problem (3.3). The optimal value of $\tilde{\boldsymbol{x}}^*$ for $\nu = 0$ is given by

$$\boldsymbol{x}^* = 0 \quad \text{and} \quad x_0^* = \max_i |\boldsymbol{b}_i| = \varphi(0)\,.$$

As $\nu$ is increased, the optimal $x_0^*(\nu)$ will decrease (as shown in Lemma 1) linearly with $\nu$ until the next basis change.

**Lemma 2:** *The total number of active bounds $x_j = \pm\nu$ is monotonically decreasing for increasing $\nu$ and goes to zero for sufficiently large $\nu$.*

Proof: For an active bound $x_j^* = \pm\nu$ in (3.2) a corresponding solution $\boldsymbol{u}^*$ exists by the KUHN-TUCKER-Theorem for the dual problem (3.3), for which the component $u_i^* > 0$ (see GOL'STEJN/JUDIN [8]), that is, we have a positive $u_i^*$ corresponding to $-\nu$ in $(\tilde{\boldsymbol{b}}, \boldsymbol{u})$. Now, if such a basis were optimal for arbitrarily large $\nu$, we would have $(\tilde{\boldsymbol{b}}, \boldsymbol{u}^*) < 0$ for sufficiently large $\nu$. But this contradicts the fact that for each fixed $\nu$ the objective function-values of (3.2) and (3.3) are equal and $(\tilde{\boldsymbol{b}}, \boldsymbol{u}^*) = (\tilde{\boldsymbol{c}}, \tilde{\boldsymbol{x}}^*) = x_0^* \geqq 0$. The monotonic decrease of the total number of active bounds of the form $x_j = \pm\nu$ follows from the convexity of $C$ and the properties of the solution-function $\varphi(\nu)$.

Suppose that we have an optimal solution $\tilde{\boldsymbol{x}}^* = (\boldsymbol{x}^*, x_0^*)$ for some fixed $\nu$, such that only one of the last $(2n+1)$ columns of $\tilde{\boldsymbol{A}}$ is in the optimal basis of (3.3). Then either $x_0^* = 0$ or $x_j^* = \pm\nu$ for one value of $i$. Furthermore, since we must have $n$ out of the first $2m$ columns of $\tilde{\boldsymbol{A}}$ in the basis, we will have $\sum_{j=1}^{n} a_{ij} x_j^* - b_i = \pm x_0^*$ for at least $n$ values of $i$. That is, the maximum error $x_0^*$ is attained for at least $n$ equations, that means, $\boldsymbol{x}^*$ gives a CHEBYSHEV error. We denote by $\underline{\nu}$ the smallest value of $\nu$, for which at most one active restriction $x_j^* = \pm\nu$ exists. For $\nu \geqq \underline{\nu}$ the solution-function $\varphi(\nu) = x_0^*(\nu)$ will reach its minimum value and we denote this value of $\nu$

by $\nu_{\min}$. For $\nu \geqq \nu_{\min}$ the quasi-solution $\boldsymbol{x}^*(\nu)$ remains constant, that is

$$\boldsymbol{x}^*(\nu) = \boldsymbol{x}^*(\nu_{\min})$$
$$x_0^*(\nu) = x_0^*(\nu_{\min}) \quad \forall \nu \geqq \nu_{\min}.$$

As the numerical results show, the CHEBYSHEV solution $\boldsymbol{x}^*(\underline{\nu})$ is closely related to the pseudoinverse solution $\boldsymbol{x}^+ = \boldsymbol{A}^+\boldsymbol{b}$ and we emphasize again that relative to $\boldsymbol{x}^+$ the quasi-solution $\boldsymbol{x}^*(\underline{\nu})$ is stable.

Now consider the case where the basismatrix in the simplex method does not change as $\nu$ increases from $\underline{\nu}$ to $\nu_{\min}$. Let us denote by $\underline{\boldsymbol{B}}$ the basis obtained when $\nu = \underline{\nu}$. $\underline{\boldsymbol{B}}$ will consist of $n$ columns from the first $2m$ columns of $\tilde{\boldsymbol{A}}$ and one column from the last $(2n+1)$ columns of $\tilde{\boldsymbol{A}}$. If this one column corresponds to the zero element in $\tilde{\boldsymbol{b}}$, then $x_0(\underline{\nu}) = 0$ and $\nu_{\min} = \underline{\nu}$. Otherwise, we have a single column in $\underline{\boldsymbol{B}}$, say the $r$-th, corresponding to one of the elements $-\nu$ in $\tilde{\boldsymbol{b}}$. Since the basis does not change, we have

$$\boldsymbol{x}(\nu) = \underline{\boldsymbol{B}}^{-1}\tilde{\boldsymbol{b}}_{\underline{\boldsymbol{B}}}(\nu), \quad \underline{\nu} \leqq \nu \leqq \nu_{\min},$$

where $\tilde{\boldsymbol{b}}_{\underline{\boldsymbol{B}}}(\nu)$ corresponds to the columns of $\underline{\boldsymbol{B}}$. Now every component of $\tilde{\boldsymbol{b}}_{\underline{\boldsymbol{B}}}(\nu)$ is constant except of the $r$-th element which is $-\nu$. Therefore

$$\boldsymbol{x}(\nu) = \boldsymbol{x}(\underline{\nu}) + (\nu - \underline{\nu})\, \boldsymbol{\beta}, \quad \underline{\nu} \leqq \nu \leqq \nu_{\min}, \tag{3.6}$$

where the constant vector $\boldsymbol{\beta}$ is the $r$-th column of $\underline{\boldsymbol{B}}^{-1}$ or the $r$-th row of $(\underline{\boldsymbol{B}}^{-1})^{\mathsf{T}}$. At $\nu = \nu_{\min}$, a new basis $\underline{\underline{\boldsymbol{B}}}$ is obtained. This new basis has no columns corresponding to the components $-\nu$ in $\tilde{\boldsymbol{b}}$. Therefore $\tilde{\boldsymbol{b}}_{\underline{\underline{\boldsymbol{B}}}}(\nu)$ is a constant vector for $\nu \geqq \nu_{\min}$. The defect $x_0(\nu)$ and the quasi-solution $\boldsymbol{x}(\nu)$ are then completely described for $\nu \geqq \underline{\nu}$:

$$\boldsymbol{x}(\nu) = \begin{cases} \boldsymbol{x}(\underline{\nu}) + (\nu - \underline{\nu})\, \boldsymbol{\beta}, & \underline{\nu} \leqq \nu \leqq \nu_{\min} \\ \boldsymbol{x}(\nu_{\min}) = \boldsymbol{x}(\underline{\nu}) + (\nu_{\min} - \underline{\nu})\, \boldsymbol{\beta}, & \nu \geqq \nu_{\min}. \end{cases}$$

It follows from the relation $x_0 \geqq |\sum a_{ij}x_j - b_i|\; i = 1(1)\,n$ and from (3.6) that $\|\boldsymbol{A}\hat{\boldsymbol{\beta}}\|_\infty \leqq \beta_0$, where $\beta_0$ is the last component of the vector $\boldsymbol{\beta}$ and $\hat{\boldsymbol{\beta}}$ represents the first $n$ components of $\boldsymbol{\beta}$, that is, $\boldsymbol{\beta} = (\hat{\boldsymbol{\beta}}, \beta_0)$. Thus the ratio $\beta_0/\|\hat{\boldsymbol{\beta}}\|_\infty$ is a good measure of the linear independence of the columns of $\boldsymbol{A}$, and will be small if the columns are almost linearly dependent. Furthermore, if this ratio is small, we can say that the vector $\hat{\boldsymbol{\beta}}$ lies in an "approximate null space" of $\boldsymbol{A}$.

## 4. Numerical Results

In order to test this approach to compute quasi-solutions of ill-conditioned equations by parametric linear programming, we have used HILBERT-matrices

$$\boldsymbol{A} = (a_{ij}) \quad \text{with} \quad a_{ij} = (i+j-1)^{-1} \quad i = 1(1)\,m, \quad j = 1(1)\,n. \tag{4.1}$$

The aim of our numerical research is the behaviour of the quasi-solutions $\boldsymbol{x}(\nu)$ for monotonic and non-monotonic solutions $\bar{\boldsymbol{x}}$ by varying the number $m$ of rows for the system matrix $\boldsymbol{A}$. In order to test the stability of the quasi-solutions $\boldsymbol{x}(\nu)$, we use different perturbed matrices $\boldsymbol{A}$ and vectors $\boldsymbol{b}$ for system (1.2). The HILBERT-matrices here used, of dimension $n=5$ and $m=7, \ldots, 10$ have in spite of small column-number $n$ a high condition. For the case $m=7$ e.g., the largest singular value is $s_1=1{,}61$ and the smallest singular value is $s_5=2{,}83 \cdot 10^{-4}$ and therefore the condition is $\operatorname{cond}(\boldsymbol{A})=5{,}67 \cdot 10^3$. The elements of $\boldsymbol{A}$ were rounded to 6 decimals. Furthermore, we have given a vector $\bar{\boldsymbol{x}}=\{\bar{x}_j\}_{j=1}^5$ and a vector $\bar{\boldsymbol{d}}=\{\bar{d}_i\}_{i=1}^m$ with $\bar{d}_i=(-1)^{i+1} \cdot 0{,}01$ for perturbation of the right-hand side, and this vector is to be reconstructed by these procedures as precisely as possible. The vector $\boldsymbol{b}_\delta$ of the perturbed righthand side is determined by $\boldsymbol{b}_\delta=\boldsymbol{A}\bar{\boldsymbol{x}}-\bar{\boldsymbol{d}}$.

As we can see in § 3, it is reasonable to use the parametric linear programming problem (3.3) for the computation of the quasi-solutions $\boldsymbol{x}(\nu)$, and problem (3.3) was computed by the computer ES 1020. For this choice of $\boldsymbol{b}_\delta$, the solution to (3.2) (vector of dual variables to problem (3.3)) is:

$$\boldsymbol{x}(\nu)=\bar{\boldsymbol{x}}; \quad x_0(\nu)=\|\boldsymbol{d}(\boldsymbol{x}(\nu))\|_\infty=0{,}01 \quad \text{for} \quad \nu \geqq \nu_{\min}.$$

In order to test the stability of these solutions, we perturbed the matrix $\boldsymbol{A}$ in such a way that to each element of $\boldsymbol{A}$ a number of the intervall $[-0{,}000005; 0{,}000005]$ was added.

### Study of Monotonically Increasing Solutions

In particular, we consider the vector $\bar{\boldsymbol{x}}=(50, 50, 100, 500, 1000)^{\mathsf{T}}$. The behaviour of the solutions $\boldsymbol{x}_\delta(\nu)$ in dependence of the parameter $\nu$ is shown in Table 1 for an unperturbed matrix $\boldsymbol{A}$ of dimension (7,5). If we perturbe the matrix $\boldsymbol{A}$, then the solutions $\boldsymbol{x}_\delta^h(\nu)$ change in such a way that we have for small parameter-values $\nu$ a small difference from the solution $\boldsymbol{x}_\delta(\nu)$ of the unperturbed problem. In particular, for the value $\nu=\underline{\nu}$, for which initially only one of the last $(2n+1)$ columns of $\tilde{\boldsymbol{A}}$ is in the basis, the difference between $\boldsymbol{x}_\delta^h(\underline{\nu})$ and $\boldsymbol{x}_\delta(\underline{\nu})$ is relatively small and for $\nu>\underline{\nu}$ this difference increases. We can see that the function $\|\boldsymbol{x}_\delta^h(\nu)-\boldsymbol{x}_\delta(\nu)\|_\infty$ at $\nu=\underline{\nu}$ takes its minimum for all perturbations (I, . . . , VI). This is shown in Figure 1. The differences between solutions $\boldsymbol{x}_\delta^h(\nu)$ and $\boldsymbol{x}_\delta(\nu)$ are compared for the parameter-values $\nu=\underline{\nu}$ and $\nu=\nu_{\min}$ in Table 2. If we increase the dimension of matrix $\boldsymbol{A}$ from (8,5) to (10,5), then the overdetermination of system (1.2) increases. In these cases, we have results similar to those, for the graph of the function $\|\boldsymbol{x}_\delta^h(\nu)-\boldsymbol{x}_\delta(\nu)\|_\infty$, for the dimension (7,5) and the computation turns out to be numerically fairly easy.

By reason of these results, we can propose for monotonically increasing solutions the quasi-solution $\boldsymbol{x}_\delta^h(\nu)$ at $\nu=\underline{\nu}$ as a stable approximation for the solution of equation (1.2). As parameter $\nu$ $(\nu \leqq \underline{\nu})$ increases, the solution $\boldsymbol{x}_\delta^h(\nu)$ tends towards the solution $\bar{\boldsymbol{x}}$; for parameter-values greater than $\underline{\nu}$ we cannot exclude (as the examples show) that the solution $\boldsymbol{x}_\delta^h(\nu)$ again moves away from $\bar{\boldsymbol{x}}$.

Table 1

Quasi-solutions $\boldsymbol{x}_\delta(\nu)$ for unperturbed matrix $\boldsymbol{A}$ of dimension (7,5) and a given vector $\overline{\boldsymbol{x}} = (50, 50, 100, 500, 1000)^{\mathrm{T}}$.

| $\nu$ | $x_0$ | $x_1$ | $x_2$ | $x_3$ | $x_4$ | $x_5$ |
|---|---|---|---|---|---|---|
| 0 | 433,3233 | 0 | 0 | 0 | 0 | 0 |
| 119,97668 | 159,37658 | 119,97668 | 119,97668 | 119,97668 | 119,97668 | 119,97668 |
| 162,38874 | 97,87915 | 162,38874 | 162,38874 | 162,38874 | 162,38874 | 162,38874 |
| 190,75975 | 66,87372 | 190,75975 | 190,75975 | 190,75975 | 190,75975 | 190,75975 |
| 211,16124 | 48,82813 | 211,16124 | 211,16124 | 211,16124 | 211,16124 | 211,16124 |
| 215,85345 | 47,65061 | 203,96205 | 215,85345 | 215,85345 | 215,85345 | 215,85345 |
| 309, 76016 | 25,03729 | 60,83516 | 309,76016 | 309,76016 | 309,76016 | 309,76016 |
| 347,22736 | 16,52431 | 4,23928 | 347,22736 | 347,22736 | 347,22736 | 347,22736 |
| 390,32139 | 7,33554 | − 60,25348 | 390,32139 | 390,32139 | 390,32139 | 390,32139 |
| 405,29627 | 5,52334 | − 92,33012 | 405,29627 | 405,29627 | 405,29627 | 405,29627 |
| 603,46562 | 0,57604 | 69,43954 | −218,81363 | 603,46562 | 603,46562 | 603,46562 |
| 624,05859 | 0,21060 | 87,29061 | −286,04119 | 624,05859 | 624,05859 | 624,05859 |
| 625,76740 | 0,19666 | 89,80437 | −292,93732 | 625,76740 | 625,76740 | 625,76740 |
| 808,22040 | 0,03149 | 52,64865 | 50,76730 | − 25,13011 | 808,22040 | 808,22040 |
| 832,66643 | 0,01279 | 47,92765 | 95,41888 | −111,00279 | 832,66643 | 832,66643 |
| 838,71347 | 0,01063 | 46,78075 | 106,44331 | −132,26820 | 838,71347 | 838,71347 |
| 842,57472 | 0 01028 | 45,71927 | 115,04900 | −147,20657 | 842,57472 | 842,57472 |
| $\underline{\nu}$=842,88962 | 0,01027 | 45,53409 | 116,03916 | −148,62304 | 842,88962 | 842,88962 |
| 990,49907 | 0,01001 | 49,55284 | 55,01761 | 83,21824 | 521,64980 | 990,49907 |
| 999,99995 − ∞ | 0,01000 | 50,00000 | 50,00002 | 99,99993 | 500,00001 | 999,99995 |

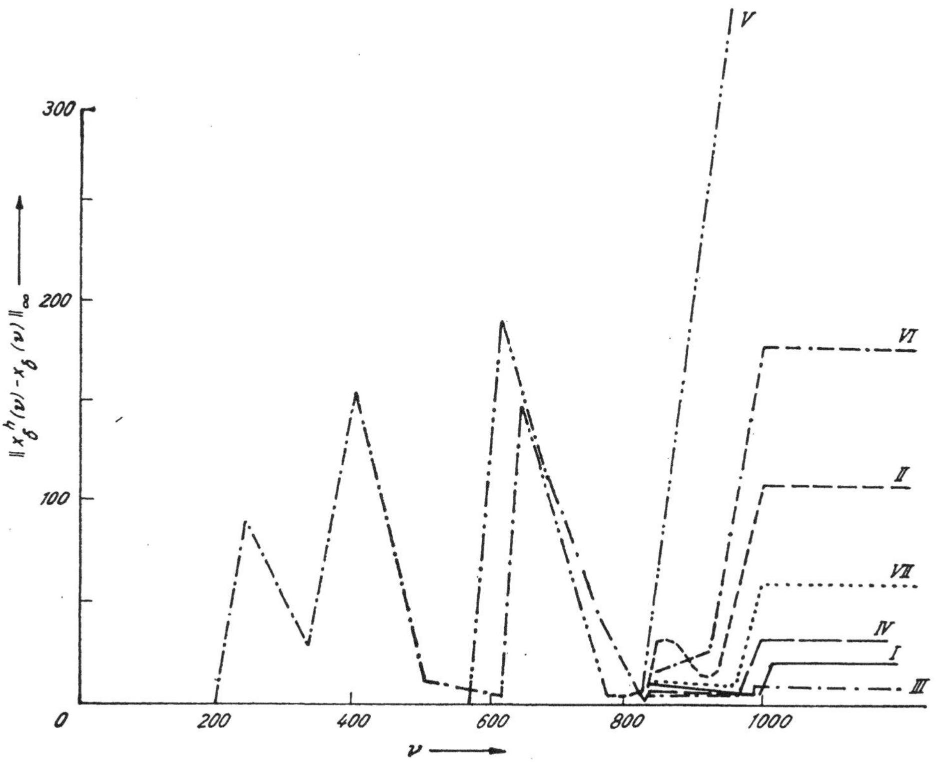


Fig. 1: Differences between perturbed quasi-solutions $x_\delta^h(\nu)$ and unperturbed quasi-solutions $x_\delta(\nu)$ for the (7,5)-matrix with $\bar{x} = (50, 50, 100, 500, 1000)^T$.

Table 2
Perturbations of the (7,5)-matrix with $\bar{x} = (50, 50, 100, 500, 1000)^T$.

| Perturbations | $\underline{\nu}$ | $\nu_{\min}$ | $\|x_\delta^h(\nu_{\min}) - x_\delta(\nu_{\min})\|$ | $\|x_\delta^h(\underline{\nu}) - x_\delta(\underline{\nu})\|$ | $\|x_\delta^h(\underline{\nu}) - \bar{x}\|$ |
|---|---|---|---|---|---|
| I | 845,28 | 1010,36 | 20,07 | 12,30 | 345,28 |
| II | 837,09 | 944,57 | 20,7 | 0,62 | 337,09 |
| III | 842,66 | 994,72 | 1,5 | 1,0 | 342,66 |
| IV | 842,27 | 985,70 | 2,9 | 2,6 | 342,27 |
| V | 836,01 | 1090,09 | 590,09 | 0,22 | 336,01 |
| VI | 839,52 | 911,48 | 25,7 | 0,9 | 339,52 |
| VII | 844,05 | 970,08 | 6,6 | 6,1 | 344,05 |

**Study of Non-monotonically and Monotonically Decreasing Solutions**

If we deal with monotonically increasing solutions, the simplexmethod for (3.3) initially contains front columns of the first $2n$ columns of $\tilde{A}$ in the basis. These columns correspond to the first rows in the HILBERT-matrix $A$ of equation system (1.2), and these columns are linearly independent, since $n$ is a small number.

If we reconstruct nonmonotonic solutions of (1.2), for example $\bar{\boldsymbol{x}} = (100, 50, 50, 1000, 500)^\mathsf{T}$, then back columns of the first $2n$ columns of $\tilde{\boldsymbol{A}}$ can be in the basis at the very beginning. Since the last columns of the HILBERT-matrix $\boldsymbol{A}^\mathsf{T}$ are nearly linearly dependent, we cannot exclude that numerical difficulties can occur in the computation of the inverse basis matrix for the simplex method. In order to prevent this case, we use procedures with repeated data-regeneration and get useful results. From numerical experience we know that also for non-monotonic solutions $\bar{\boldsymbol{x}}$ the quasi-solutions $\boldsymbol{x}_\delta^h(\nu)$ at $\nu = \underline{\nu}$ are relatively stable regarding perturbations in the matrix $\boldsymbol{A}$ (see Fig. 2 and Table 3). The differences between $\boldsymbol{x}_\delta^h(\underline{\nu})$ and $\bar{\boldsymbol{x}}$ are relatively small, and so we can use $\boldsymbol{x}_\delta^h(\underline{\nu})$ as an approximate solution for equations (1.2).

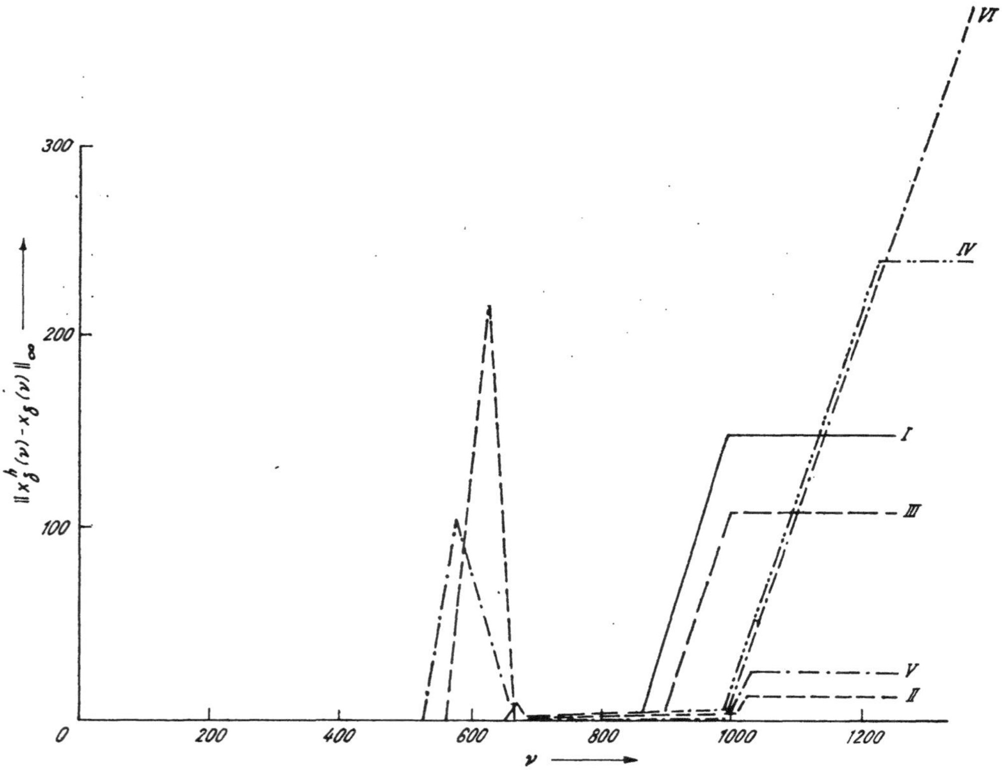


Fig. 2: Differences between perturbed quasi-solutions $\boldsymbol{x}_\delta^h(\nu)$ and unperturbed quasi-solutions $\boldsymbol{x}_\delta(\nu)$ for the (7,5)-matrix with $\bar{\boldsymbol{x}} = (100, 50, 50, 1000, 500)^\mathsf{T}$.

The investigation of monotonically decreasing solutions, for instance $\bar{\boldsymbol{x}} = (1000, 500, 100, 50, 50)^\mathsf{T}$ shows great variations of solutions $x_\delta^h(\nu)$, and the distance between $\boldsymbol{x}_\delta^h(\underline{\nu})$ and $\bar{\boldsymbol{x}}$ is very large. However, we can see that these distances at the parameter-value $\nu = \underline{\nu}$ are relatively small (see Fig. 3 and Table 4). But by reason of the large distance between $\boldsymbol{x}_\delta^h(\underline{\nu})$ and $\bar{\boldsymbol{x}}$, the use of solutions $\boldsymbol{x}_\delta^h(\underline{\nu})$ as approximate

Table 3
Perturbations of the (7,5)-matrix with $\bar{x} = (100, 50, 50, 1000, 500)^T$.

| Perturbations | $\underline{\nu}$ | $\nu_{\min}$ | $\Vert x_\delta^h(\nu_{\min}) - x_\delta(\nu_{\min})\Vert$ | $\Vert x_\delta^h(\underline{\nu}) - x_\delta(\underline{\nu})\Vert$ | $\Vert x_\delta^h(\underline{\nu}) - \bar{x}\Vert$ |
|---|---|---|---|---|---|
| I | 661,60 | 851,23 | 0,9 | 0,1 | 338,4 |
| II | 661,45 | 1012,47 | 12,47 | 1,4 | 338,5 |
| III | 661,83 | 892,34 | 0,6 | 0,9 | 338,2 |
| IV | 659,84 | 1238,57 | 238,57 | 0,5 | 340,2 |
| V | 663,41 | 1027,66 | 27,66 | 6,6 | 336,6 |
| VI | 662,11 | 2077,45 | 1077,45 | 1,8 | 337,9 |

solutions for the equation system (1.2) is questionable. These results also apply for dimensions $m = 8, \ldots, 10$ and $n = 5$.

We can therefore state that by using the $L_\infty$-norm and the compact $C$ the problem (3.1) of minimizing the defect vector of an overdetermined, ill-conditioned equation system over this family of compacts $C$ results in a problem of linear parametric optimization. This problem is solvable by the well-known simplex methods. We obtain quasi-solutions $x_\delta(\nu)$ for problem (1.2) in dependence of the parameter $\nu$,

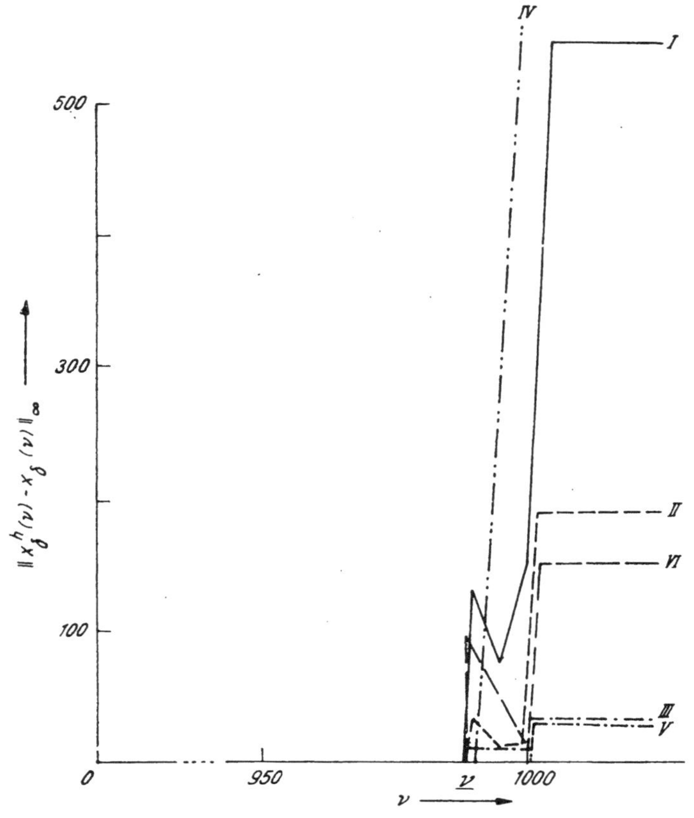


Fig. 3: Differences between perturbed quasi-solutions $x_\delta^h(\nu)$ and unperturbed quasi-solutions $x_\delta(\nu)$ for the (7,5)-matrix with $\bar{x} = (1000, 500, 100, 50, 50)^T$.

Table 4
Perturbations of the (7,5)-matrix with $\bar{x}=(1000, 500, 100, 50, 50)^T$.

| Perturbations | $\underline{\nu}$ | $\nu_{min}$ | $\lVert x_\delta^h(\nu_{min}) - x_\delta(\nu_{min})\rVert$ | $\lVert x_\delta^h(\underline{\nu}) - x_\delta(\underline{\nu})\rVert$ | $\lVert x_\delta^h(\underline{\nu}) - \bar{x}\rVert$ |
|---|---|---|---|---|---|
| I | 988,24 | 1005,41 | 546,31 | 0,1 | 938,24 |
| II | 989,28 | 1001,66 | 190,1 | 0,3 | 939,28 |
| III | 989,61 | 999,53 | 6,1 | 5,1 | 939,61 |
| IV | 989,34 | 1346,36 | 1296,36 | 1,2 | 939,34 |
| V | 989,94 | 999,50 | 2,2 | 38,3 | 939,94 |
| VI | 988,57 | 1001,54 | 151,1 | 0,2 | 938,57 |

which is the diameter of the compact $C$. Numerical experience provides the following results: for monotonically increasing and non-monotonically solutions of a system (1.2) we get quasi-solutions $x_\delta^h(\nu)$ (CHEBYSHEV-solutions) which are relatively stable for small perturbations in the data of matrix $A$ at the uniquely determined parameter-value $\nu=\underline{\nu}$. These we can use as stable approximate solutions for the equation system. For systems with monotonically decreasing solutions we too get stable quasi-solutions $x_\delta^h(\nu)$ for parameter-value $\nu=\underline{\nu}$, but these are very different from the solution $\bar{x}$. Therefore they are unsuitable as approximate solutions.

An interpretation of the different error-behaviour of these quasi-solutions for the various vectors $\bar{x}$ can be made, if we decompose the solution-vector by eigenvectors $\bar{x}=\sum(\bar{x}, u_i)\, u_i$ of the system matrix $A^TA$ (see (1.6) and Table 5). While for $\bar{x}=(50, 50, 100, 500, 1000)^T$ and $\bar{x}=(100, 50, 50, 1000, 500)^T$ all decomposition coefficients $(\bar{x}, u_i)$ are located in the same order $10^2$ of absolute values, the coefficient $(\bar{x}, u_1)=1{,}03 \cdot 10^3$ of the greatest singular value $s_1$ for $\bar{x}=(1000, 500, 100, 50, 50)^T$ dominates the coefficients $(\bar{x}, u_2), \ldots, (\bar{x}, u_5)$ essentially, which are one or two orders smaller. For such situation we get a fairly exact reconstruction of this coefficient $(\bar{x}, u_1)$ by the quasi-solution $x(\underline{\nu})$. In the direction of the other eigenvectors $u_i$ $i=2(1)\,5$ extremely large differences may occur, which can have relative reconstruction-errors at $100\%$, i.e.

$$\frac{\lVert x(\underline{\nu})-\bar{x}\rVert_\infty}{\lVert\bar{x}\rVert_\infty}\approx 1\ .$$

Table 5
Decomposition of the vector $\bar{x}$, which is to reconstruct, in eigenvectors $u_i$ for the singular values $s_i$ of the matrix $A$ with $m=7$, $n=5$:

| $\bar{x}$ | monotonically increasing[1] | non monotonically[2] | monotonically decreasing[3] | singular values |
|---|---|---|---|---|
| $(x, u_1)$ | $4{,}42 \cdot 10^2$ | $4{,}85 \cdot 10^2$ | $1{,}03 \cdot 10^3$ | $s_1=1{,}61 \cdot 10^0$ |
| $(x, u_2)$ | $6{,}86 \cdot 10^2$ | $6{,}38 \cdot 10^2$ | $-3{,}98 \cdot 10^2$ | $s_2=2{,}37 \cdot 10^{-1}$ |
| $(x, u_3)$ | $6{,}75 \cdot 10^2$ | $5{,}55 \cdot 10^2$ | $-1{,}05 \cdot 10^2$ | $s_3=1{,}54 \cdot 10^{-2}$ |
| $(x, u_4)$ | $3{,}29 \cdot 10^2$ | $-3{,}07 \cdot 10^2$ | $8{,}82 \cdot 10^1$ | $s_4=5{,}74 \cdot 10^{-4}$ |
| $(x, u_5)$ | $1{,}83 \cdot 10^2$ | $4{,}67 \cdot 10^2$ | $8{,}66 \cdot 10^1$ | $s_5=2{,}83 \cdot 10^{-4}$ |

[1] $\bar{x}=(50, 50, 100, 5000, 1000)^T$
[2] $\bar{x}=(100, 50, 50, 1000, 500)^T$
[3] $\bar{x}=(1000, 500, 100, 50, 50)^T$

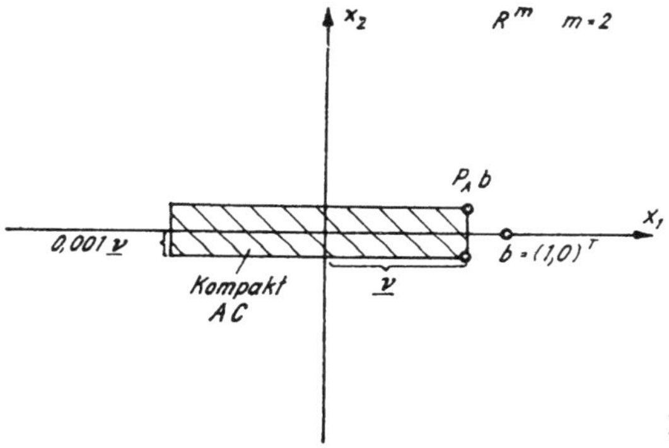


Fig. 4

The following small experiment with $m=n=2$ demonstrates this case (see Fig. 4): we consider the matrix

$$A=\begin{pmatrix}1 & 0\\ 0 & 0{,}001\end{pmatrix},\quad \bar{x}=(1,\,0)^{\mathsf{T}}=u_1,\quad u_2=(0,\,1)^{\mathsf{T}},\quad s_1=1,\quad s_2=0{,}001$$

and the system $A\bar{x}=(1,\,0)^{\mathsf{T}}=b$ for an unperturbed right-hand side. Then holds for $\underline{\nu}$ (as in Fig. 4 shown): $1-\underline{\nu}=0{,}001\cdot\underline{\nu}$, and we have therefore

$$\underline{\nu}=\frac{1000}{1001}\quad\text{with}\quad \mathsf{P}_A b=\left(\frac{1000}{1001};\ \pm\frac{1}{1001}\right)^{\mathsf{T}}\quad\text{and}$$

$$x(\underline{\nu})=\left(\frac{1000}{1001};\ \pm\frac{1000}{1001}\right)^{\mathsf{T}}\quad\text{and this leads to}\quad \|x(\underline{\nu})-\bar{x}\|_\infty=\frac{1000}{1001}\approx 1\,.$$

The accuracy $\|x(\underline{\nu})-\bar{x}\|_\infty$ is in general better for solutions $\bar{x}$, where all decomposition coefficients are located in the same order, because the reconstruction-error is distributed over all solution components. It is interesting, that we get good approximations with $L_\infty$-quasi-solutions also in those cases where EUCLIDean quasi-solutions or regularized solutions equivalent to them are very unprecise, because strong damping has taken plcae, in particular for such decomposition coefficients $(\bar{x},\,u_i)$, which correspond to small singular values $s_i$. On the other hand, as these examples have shown, we can see that by TYKHONOV's regularization approximated solutions are very reliable in cases with dominant portions for large singular values, where $L_\infty$-norm is unfavourable.

**References**


[1] ALIEV, B.: Eine Abschätzung der Regularisierungsmethode für algebraische Gleichungen (russ.). DAN Tadz. SSR **19** (1976) 10, 3–5.

[2] ANTON, H.; DURIS, C. S.: On minimum norm and best approximate solutions of $Av=b$ in normed spaces. J. Approx. Theory **16** (1976) 3, 245–250.

[3] BARTELS, R. H.; GOLUB, G. H.: Stable numerical methods for obtaining the Chebyshev solution to an overdetermined system of equations. Comm. ACM **11** (1968) 403–408.

[4] Chavent, G.: About the stability of the optimal control solution of inverse problems. In: Inverse and improperty posed problems in differential equations, ed. by G. Anger, Akad. Verlag Berlin 1979.

[5] Dieudonné, J.: Grundzüge der modernen Analysis. Verlag d. Wiss., Berlin 1971.

[6] Forshyte, G. E.; Moler, C. B.: Computer-Verfahren für lineare algebraische Systeme, Oldenburg-Verlag München 1971.

[7] Friedrich, V.; Hofmann, B.; Tautenhahn, U.: Möglichkeiten der Regularisierung bei der Auswertung von Meßdaten. WSR THK 1979.

[8] Gol'stejn, E. G.; Judin, D. B.: Neue Richtungen in der linearen Optimierung (russ.). Moskau 1966.

[9] Ivanov, W. K.: Über lineare nichtkorrekte Aufgaben (russ.). DAN SSSR **145** (1962) 2, 270–272.

[10] Ivanov, W. K.: Über nichtkorrekt gestellte Aufgaben (russ.). Mat. Sbornik **61** (1963) 2, 211–223.

[11] Kuhnert, F.: Pseudoinverse Matrizen und die Methode der Regularisierung. Teubner Verlag Leipzig 1976.

[12] Nožička, F.; Guddat, J.; Hollatz, H.; Bank, B.: Theorie der linearen parametrischen Optimierung, Akad. Verlag Berlin 1974.

[13] Rosen, J. B.: Chebyshev solution of large linear systems. J. Computer and System Science **1** (1967) 29–43.

[14] Tichonov, A. N.; Arsenin, W. J.: Methoden der Lösung nichtkorrekter Aufgaben (russ.). Moskau 1974.

**Zusammenfassung**

Gegenstand der Untersuchungen ist die Stabilität von $L_\infty$-Quasilösungen im Sinne von Ivanov für bestimmte, schlecht konditionierte lineare Gleichungssysteme. Es wird ein Stabilitätssatz bewiesen, der auch Störungen in der Koeffizientenmatrix des Gleichungssystems berücksichtigt. Mit Hilfe der linearen parametrischen Optimierung wird anschließend ein effektiver Zugang zur Berechnung von Quasilösungen im Sinne der $L_\infty$-Norm aufgezeigt. Die angeführten numerischen Untersuchungen gestatten Aussagen zum Stabilitätsverhalten derartiger Lösungen bei gestörter rechter Seite und gestörter Matrix des Gleichungssystems. Es erweist sich, daß mit $L_\infty$-Quasilösungen gerade dann gute Näherungen erreicht werden können, wenn Euklidische Quasilösungen oder äquivalente regularisierte Lösungen sehr ungenau sind.

**Резюме**

Изучена проблема устойчивости $L_\infty$-квази-решений в смысле В. К. Иванова для переопределенных, плохо обусловленных линейных систем уравнений. Доказана теорема устойчивости, в которой учитываются нарушения и в матрице коэффициентов и в правой части системы уравнений. На базе линейного параметрического программирования описывается эффективный метод определения $L_\infty$-квази-решений. Проведенные вычислительные эксперименты дают показания о поведении таких решений при нарушениях правой части и матрицы системы уравнений. Если евклидовые квази-решения или эквивалентные регуляризированные решений являются достаточно неточными, то с помощью $L_\infty$-квази-решений получаются довольно хорошие аппроксимации.

Received November 1979

Math. Operationsforsch. Statist., Ser. Optimization, Vol. 11 (1980) No. 4, 579–591

# Задача убегания от группы преследователей

Е. Ф. Мищенко[1], М. С. Никольский[2], Н. Сатимов[3]

В статье развиваются методы решения задачи уклонения от встречи одного убегающего объекта от нескольких преследующих объектов. Эффективность полученных ресультатов продемонстрирована на ряде примеров общего вида. В статье также получены оценки снизу для расстояния от убегающего объекта до каждого из догоняющих объектов в процессе уклонения от встречи.

## 0. Введение

Пример задачи убегания от группы преследователей дает следующий процесс уклонения одного управляемого объекта от группы других управляемых объектов. Точки $\boldsymbol{x}_1, \ldots, \boldsymbol{x}_r$, принадлежащие фазовому пространству $R^s$, преследуют точку $\boldsymbol{y}$, принадлежащую тому же пространству. Движения точек описываются соответственно дифференциальными уравнениями

$$\dot{\boldsymbol{x}}_i = f_i(\boldsymbol{x}_1, \ldots, \boldsymbol{x}_r, \boldsymbol{u}_i), \quad i = 1, \ldots, r; \quad \dot{\boldsymbol{y}} = g(\boldsymbol{y}, \boldsymbol{v}) .$$

В этих уравнениях $\boldsymbol{u}_i$ – управляющий вектор $i$-го преследующего объекта, $\boldsymbol{v}$ – управляющий вектор убегающего объекта. Точка $\boldsymbol{y}$ считается пойманной, если по крайней мере для одного значения индекса $i$, $i = 1, \ldots, r$, определенная часть координат вектора $\boldsymbol{x}$ совпадает с такой же частью координат вектора $\boldsymbol{y}$. Цель убегающего – избежать поимки.

Из описанного процесса уклонения мы получим дифференциальную игру, если введем новый фазовый вектор $\boldsymbol{z} = (\boldsymbol{x}_1, \ldots, \boldsymbol{x}_r, \boldsymbol{y})$ прямого произведения $R^{(r+1)s}$ фазовых пространств векторов $\boldsymbol{x}_1, \ldots, \boldsymbol{x}_r, \boldsymbol{y}$ и запишем уравнения в виде одного дифференциального уравнения

$$\dot{\boldsymbol{z}} = F(\boldsymbol{z}, \boldsymbol{u}, \boldsymbol{v}) ,$$

где $\boldsymbol{u} = (\boldsymbol{u}_1, \ldots, \boldsymbol{u}_r)$.

Условия поимки точки $\boldsymbol{y}$ точками $\boldsymbol{x}_1, \ldots, \boldsymbol{x}_r$ выделяют в пространстве $R^{(r+1)s}$ терминальное множество $M$, представляющее собой объединение $r$ линейных подпространств.

---

[1,2] Математический институт АН СССР, Москва, СССР 117333 Москва Вавилова 42.
[3] ТГУ Ташкент, факультет прикладной математики и механики. СССР 700095 Ташкент.

## 1. Постановка задачи и формулировки основных результатов

**А.** В настоящей работе рассматриваются квазилинейные дифференциальные игры, описываемые уравнениями вида

$$\dot{z} = Cz + f(u, v)\ , \tag{1}$$

где $z \in R^n$, $R^n$ — $n$-мерное векторное евклидово пространство, $C$ — линейное отображение $R^n$ в себя, $u, v$ — управляющие векторы, $u$ — вектор преследования, $v$ — вектор убегания, $u \in P$, $v \in Q$, $P$ и $Q$ — непустые компактные подмножества пространств $R^p$ и $R^q$ соответственно, $f$ — непрерывное отображение множества $P \times Q$ в $R^n$. В пространстве $R^n$ выделено терминальное множество $M$, представляющее собой объединение $\bigcup_{i=1}^{m} M_i$ линейных подпространств $M_i \subset R^n$.

Будем говорить, что из точки $z_0 \in R^n \setminus M$ можно уклониться от встречи с множеством $M$, если по любой измеримой функции $u(t)$, $0 \leqq t < \infty$, $u(t) \in P$, можно построить такое измеримое изменение $v(t)$ вектора $v \in Q$, что решение $z(t)$ уравнения

$$\dot{z} = Cz + f(u(t), v(t)), \quad z(0) = z_0\ , \tag{2}$$

ни при каком значении времени $t$, $0 < t < \infty$, не попадает на множество $M: z(t) \notin M$ для всех $t \in (0, \infty)$. При этом для нахождения значения $v(t)$ вектора $v$ в каждый момент времени $t \geqq 0$ разрешается использовать лишь значения $u(t)$ вектора $u$ в тот же момент времени $t$ и $z(s)$ вектора $z$ при $0 \leqq s \leqq t$.

В том случае, когда можно уклониться от встречи с множеством $M$ из произвольной точки $z_0 \in R^n \setminus M$, мы будем говорить, что в игре (1) возможно убегание.

Задаче убегания для различных классов дифференциальных игр посвящено много работ (см., например, [1]–[19]). В настоящей работе получены некоторые новые достаточные условия для возможности убегания и $l$ — убегания в игре (1). Работа примыкает естественным образом к работам Е. Ф. Мищенко, Н. Сатимова [7] и Е. Ф. Мищенко, М. С. Никольского, Н. Сатимова [8].

**Б.** Введем некоторые обозначения. $R^r$ — $r$-мерное евклидово арифметическое пространство, $L_i$ — прямое дополнение подпространства $M_i$ в $R^n$, $\xi_i(\eta_i)$ — расстояние от точки $z$ до $M_i(L_i)$, $\xi_i(t)$ $(\eta_i(t))$ — расстояние от точки $z(t)$ до $M_i(L_i)$, $W_i$ и $\tilde{W}_i$ — подпространства из $L_i$ такие, что $L_i = W_i \oplus \tilde{W}_i$, $\Pi_i$ — оператор проектирования из $R^n$ на $L_i$ параллельно $M_i$, $\Delta_i$ — оператор проектирования из $L_i$ на $W_i$ параллельно $\tilde{W}_i$, $\pi_i = \Delta_i \Pi_i$, $0$ — нулевой вектор пространства $R^r$, $S$ — единичная сфера $R^\nu$ с центром в точке $0$, индекс $i$ принимает значения $1, \ldots, m$, $(z_1, z_2)$ — скалярное произведение векторов $z_1, z_2 \in R^r$, константы — положительные числа, зависящие лишь от игры (1), но не зависящие ни от начальной точки $z_0 \in R^n \setminus M$, ни от измеримых управляющих функций $u(t)$, $v(t)$.

Пусть $\mu>0$. Обозначим через $M_\mu$ совокупность всех точек $z\in R^n$ таких, что расстояние от $z$ до $M$ положительно и $\leqq\mu$.

**B**. Формулировки предположений и основных результатов.

**Предположение 1:** Пусть для игры (1) существуют натуральные числа $k_i$ и $\nu$, подпространства $L_i\subset R^n$, число $\mu>0$, векторы $\boldsymbol{l}, \boldsymbol{l}_i\in R^\nu$, линейные отображения $F_i\colon L_i\to R^\nu$ и непрерывные отображения $\Phi\colon P\times Q\to R^\nu$, $\Phi_i\colon P\times Q\to R^\nu$ такие, что: а) каждое множество $F_i\Pi_iC^jf(P, Q), j=0, \ldots, k_i-2$, является одноточечным; б) $F_i\Pi_iC^{k_i-1}f(\boldsymbol{u}, \boldsymbol{v})=\Phi(\boldsymbol{u}, \boldsymbol{v})+\Phi_i(\boldsymbol{u}, \boldsymbol{v})$ для всех $(\boldsymbol{u}, \boldsymbol{v})\in P\times Q$, причем по крайней мере одно из чисел $N_1-N_3$, $N_2-N_3$ положительно, где

$$N_1=\inf_{z_0\in M_\mu}\max_{\Psi\in D(z_0)}\max_{\boldsymbol{v}\in Q}\min_{\boldsymbol{u}\in P}(\Psi, \Phi(\boldsymbol{u}, \boldsymbol{v})-\boldsymbol{l}),$$

$$N_2=\inf_{z_0\in M_\mu}\max_{\Psi\in D(z_0)}\min_{\boldsymbol{u}\in P}\max_{\boldsymbol{v}\in Q}(\Psi, \Phi(\boldsymbol{u}, \boldsymbol{v})-\boldsymbol{l}),$$

$$N_3=\max_{\substack{i\\ \boldsymbol{u}\in P, \boldsymbol{v}\in Q}}|\Phi_i(\boldsymbol{u}, \boldsymbol{v})-\boldsymbol{l}_i|,$$

множество $D(z_0)\subset R^\nu$ состоит из векторов $\Psi\in S\subset R^\nu$, удовлетворяющих неравенствам: $(\Psi, F_i\Pi_iz_0)\geqq 0$,

$$(\Psi, F_i\Pi_iCz_0+F_i\Pi_if(P, Q))\geqq 0, \ldots,$$

$$(\Psi, F_i\Pi_iC^{k_i-1}z_0+F_i\Pi_iC^{k_i-2}f(P, Q))\geqq 0,$$

$(\Psi, F_i\Pi_iC^{k_i}z_0+\boldsymbol{l}+\boldsymbol{l}_i)\geqq 0$ при всех $i=1, \ldots, m$, оно предполагается непустым для всех $z\in M_\mu$.

**Теорема 1:** *Если выполнено предположение 1, то в игре (1) возможно убегание. При этом процессом уклонения от встречи с множеством $M$ из точки $z_0\in R^n\setminus M$ можно управлять так, что при всех $t\geqq 0$ будет верна оценка*

$$\xi_i(t)>\begin{cases}\dfrac{c\xi_i^k(0)}{[1+|z(t)|]^{k^2+k}}, & \text{если}\quad \xi_i(0)<\dfrac{\varepsilon}{[1+|z_0|]^k},\\[2ex] \dfrac{c\varepsilon^k}{[1+|z(t)|]^{k^2+k}}, & \text{если}\quad \xi_i(0)\geqq\dfrac{\varepsilon}{[1+|z_0|]^k},\end{cases}\tag{3}$$

*где $c$, $k$ и $\varepsilon$ — константы. В случае, когда $N_1-N_3>0$, для нахождения $\boldsymbol{v}(t)$ достаточно знания $z(s)$ при $0\leqq s\leqq t$.*

**Предположение 2:** Пусть для игры (1) существуют натуральные числа $k_i$ и $\nu$, подпространства $L_i\subset R^n$, число $\vartheta>0$, векторы $\boldsymbol{l}, \boldsymbol{l}_i\in R^\nu$, линейные отображения $F_i\colon L_i\to R^\nu$ и непрерывные отображения $\Phi\colon P\times Q\to R^\nu$, $\Phi_i\colon P\times\times Q\to R^\nu$ такие, что: а) каждое множество $F_i\Pi_iC^jf(P, Q), j=0, \ldots, k_i-2$, является одноточечным; б) $F_i\Pi_iC^{k_i-1}f(\boldsymbol{u}, \boldsymbol{v})=\Phi(\boldsymbol{u}, \boldsymbol{v})+\Phi_i(\boldsymbol{u}, \boldsymbol{v})$ для всех $(\boldsymbol{u}, \boldsymbol{v})\in P\times Q$, причем по крайней мере одно из чисел $N_1-N_3$, $N_2-N_3$ положительно, где

$$N_1 = \inf_{z_0 \in R^n \setminus M} \max_{\psi \in E(z_0)} \max_{v \in Q} \min_{u \in P} (\Psi, \Phi(u, v) - l),$$

$$N_2 = \inf_{z_0 \in R^n \setminus M} \max_{\psi \in E(z_0)} \min_{u \in P} \max_{v \in Q} (\Psi, \Phi(u, v) - l),$$

$$N_3 = \max_{\substack{i \\ u \in P, v \in Q}} |\Phi_i(u, v) - l_i|,$$

множество $E(z_0) \subset R^\nu$ состоит из векторов $\Psi \in S \subset R^\nu$, удовлетворяющих неравенствам $\left(\Psi, F_i \Pi_i e^{tC} z_0 + t F_i \Pi_i f(P, Q) + \ldots + \frac{t^{k_i-1}}{(k_i-1)!} F_i \Pi_i C^{k_i-2} f(P, Q) + \frac{t^{k_i}}{k_i!}(l + l_i)\right) \geqq$ $\geqq 0$ для всех $t \in [0, 2\vartheta]$, $i = 1, \ldots, m$, оно предполагается непустым для всех $z_0 \in R^n \setminus M$.

**Теорема 2:** *Если выполнено предположение 2, то в игре (1) возможно убегание. При этом процессом уклонения от встречи с множеством $M$ из точки $z_0 \in R^n \setminus M$ можно управлять так, что*

$$\xi_i(t) \geqq \begin{cases} c_i t^{k_i}, & \text{если} \quad 0 \leqq t < \vartheta_1, \\ l_i^0, & \text{если} \quad t \geqq \vartheta_1, \end{cases} \tag{4}$$

*где $c_i$, $\vartheta_1$ и $l_i^0$ – константы. В случае, когда $N_1 - N_3 > 0$, для нахождения значения $v(t)$ достаточно знания $z(s)$ при $0 \leqq s \leqq t$.*

**Предположение 3:** Пусть для игры (1) существуют натуральные числа $k_i$, подпространства $L_i \subset R^n$, подпространства $W_i$, $\tilde{W}_i$ из $L_i$, число $\mu > 0$ и векторы $l_i \in W_i$ такие, что: а) каждое множество $\pi_i C^j f(P, Q)$, $j = 0, \ldots, k_i - 2$, является одноточечным; б) по крайней мере одно из чисел $\tilde{N}_1$, $\tilde{N}_2$ положительно, где

$$\tilde{N}_1 = \inf_{z_0 \in M_\mu} \max_{\substack{\psi_1 \in D_1(z_0) \\ \cdots\cdots\cdots \\ \psi_m \in D_m(z_0)}} \max_{v \in Q} \min_{u \in P} \min_{i} (\Psi_i, \pi_i C^{k_i-1} f(u, v) - l_i),$$

$$\tilde{N}_2 = \inf_{z_0 \in M_\mu} \max_{\substack{\psi_1 \in D_1(z_0) \\ \cdots\cdots\cdots \\ \psi_m \in D_m(z_0)}} \min_{u \in P} \max_{v \in Q} \min_{i} (\Psi_i, \pi_j C^{k_i-1} f(u, v) - l_i),$$

множество $D_i(z_0) \subset W_i$ состоит из векторов $\Psi \in W_i$, $|\Psi| = 1$, удовлетворяющих неравенствам:

$$\begin{gathered} (\Psi, \pi_i z_0) \geqq 0, \quad (\Psi, \pi_i C z_0 + \pi_i f(P, Q)) \geqq 0, \ldots, \\ (\Psi, \pi_i C^{k_i-1} z_0 + \pi_i C^{k_i-2} f(P, Q)) \geqq 0, \quad (\Psi, \pi_i C^{k_i} z_0 + l_i) \geqq 0, \end{gathered} \tag{5}$$

оно предполагается непустым $\forall z_0 \in M_\mu$.

**Теорема 3:** *Если выполнено предположение 3, то в игре (1) возможно убегание. При этом процессом уклонения от встречи с множеством $M$ из точки $z_0 \in R^n \setminus M$ можно управлять так, что при всех $t \geqq 0$ будет верна оценка (3). В случае, когда $\tilde{N}_1 > 0$, для построения управления убегания $v(t)$ достаточно использовать знание $z(s)$, $0 \leqq s \leqq t$.*

**Предположение 4:** Пусть для игры (1) существуют натуральные числа $k_i$, подпространства $L_i \subset R^n$, подпространства $W_i$, $\tilde{W}_i$ из $L_i$, положительные числа $\vartheta$, $\mu$ и векторы $\boldsymbol{l}_i \in W_i$ такие, что: а) каждое множество $\pi_i \boldsymbol{C}^j f(P, Q)$, $j=0, \ldots, k_i-2$, является одноточечным; б) по крайней мере одно из чисел $\tilde{\tilde{N}}_1$, $\tilde{\tilde{N}}_2$ положительно, где

$$\tilde{\tilde{N}}_1 = \inf_{\boldsymbol{z}_0 \in M_\mu} \max_{\substack{\Psi_1 \in E_1(\boldsymbol{z}_0) \\ \cdots\cdots \\ \Psi_m \in E_m(\boldsymbol{z}_0)}} \max_{\boldsymbol{v} \in Q} \min_{\boldsymbol{u} \in P} \min_i \left(\Psi_i, \pi_i \boldsymbol{C}^{k_i-1} f(\boldsymbol{u}, \boldsymbol{v}) - \boldsymbol{l}_i\right),$$

$$\tilde{\tilde{N}}_2 = \inf_{\boldsymbol{z}_0 \in M_\mu} \max_{\substack{\Psi_1 \in E_1(\boldsymbol{z}_0) \\ \cdots\cdots \\ \Psi_m \in E_m(\boldsymbol{z}_0)}} \min_{\boldsymbol{u} \in P} \max_{\boldsymbol{v} \in Q} \min_i \left(\Psi_i, \pi_i \boldsymbol{C}^{k_i-1} f(\boldsymbol{u}, \boldsymbol{v}) - \boldsymbol{l}_i\right),$$

множество $E_i(\boldsymbol{z}_0) \subset W_i$ состоит из векторов $\Psi \in W_i$, $|\Psi|=1$, удовлетворяющих неравенству

$$\left(\Psi, \pi_i e^{t\boldsymbol{C}} \boldsymbol{z}_0 + \sum_{j=0}^{k_i-2} \frac{t^{j+1}}{(j+1)!} \pi_i \boldsymbol{C}^j f(P, Q) + \frac{t^{k_i}}{k_i!} \boldsymbol{l}_i\right) \geqq 0 \tag{6}$$

для всех $t \in [0, \vartheta]$, оно предполагается непустым $\forall \boldsymbol{z}_0 \in M_\mu$.

**Теорема 4:** *Если выполнено предположение 4, то в игре (1) возможно убегание. При этом процессом уклонения от встречи с множеством $M$ из точки $\boldsymbol{z}_0 \in R^n \setminus M$ можно управлять так, что при всех $t \geqq 0$ будет верна оценка:*

$$\xi_i(t) > \begin{cases} \dfrac{c\xi_i^k(0)}{[1+\eta_i(t)]^k}, & \text{если} \quad \xi_i(0) < \varepsilon, \\ \dfrac{c\varepsilon^k}{[1+\eta_i(t)]^k}, & \text{если} \quad \xi_i(0) \geqq \varepsilon, \end{cases} \tag{7}$$

*где $c$, $k$ и $\varepsilon$ — константы. В случае, когда $\tilde{\tilde{N}}_1 > 0$, для построения управления убегания $\boldsymbol{v}(t)$ достаточно использовать знание $\boldsymbol{z}(s)$, $0 \leqq s \leqq t$.*

**Предположение 5:** Пусть для игры (1) существуют натуральные числа $k_i$, подпространства $L_i \subset R^n$, $W_i$, $\tilde{W}_i$ из $L_i$, число $\vartheta > 0$ и векторы $\boldsymbol{l}_i \in W_i$ такие, что: а) каждое множество $\pi_i \boldsymbol{C}^j f(P, Q)$, $j=0, \ldots, k_i-2$, является одноточечным; б) по крайней мере одно из чисел $N_1^*$, $N_2^*$ положительно, где

$$N_1^* = \inf_{\boldsymbol{z}_0 \in R^n \setminus M} \max_{\substack{\Psi_1 \in E_1(\boldsymbol{z}_0) \\ \cdots\cdots \\ \Psi_m \in E_m(\boldsymbol{z}_0)}} \max_{\boldsymbol{v} \in Q} \min_{\boldsymbol{u} \in P} \min_i \left(\Psi_i, \pi_i \boldsymbol{C}^{k_i-1} f(\boldsymbol{u}, \boldsymbol{v}) - \boldsymbol{l}_i\right),$$

$$N_2^* = \inf_{\boldsymbol{z}_0 \in R^n \setminus M} \max_{\substack{\Psi_1 \in E_1(\boldsymbol{z}_0) \\ \cdots\cdots \\ \Psi_m \in E_m(\boldsymbol{z}_0)}} \min_{\boldsymbol{u} \in P} \max_{\boldsymbol{v} \in Q} \min_i \left(\Psi_i, \pi_i \boldsymbol{C}^{k_i-1} f(\boldsymbol{u}, \boldsymbol{v}) - \boldsymbol{l}_i\right),$$

множество $E_i(\boldsymbol{z}_0) \subset W_i$, состоит из векторов $\Psi \in W_i$, $|\Psi|=1$, удовлетворяющих неравенству

$$\left(\Psi,\ \pi_i e^{tC} z_0 + \sum_{j=0}^{k_i-2} \frac{t^{j+1}}{(j+1)!}\pi_i C^j f(P,\ Q) + \frac{t^{k_i}}{k_i!} l_i\right) \geqq 0$$

для всех $t \in [0,\ 2\vartheta]$, оно предполагается непустым $\forall z_0 \in R^n \backslash M$.

**Теорема 5:** *Если выполнено предположение 5, то в игре (1) возможно убегание. При этом процессом уклонения от встречи с множеством $M$ из точки $z_0 \in R^n \backslash M$ можно управлять так, что при всех $t \geqq 0$ будет верна оценка (4). В случае, когда $N_1^* > 0$, для построения управления убегания $v(t)$ достаточно использовать знание $z(s)$, $0 \leqq s \leqq t$.*

В 2 изложены доказательства сформулированных выше теорем, в 3 рассмотрены примеры.

## 2. Доказательства теорем

В пунктах **А**, **Б** доказана теорема 1, в пункте **В** — теорема 2, в пункте **Г** — теоремы 3–5. Ниже используются некоторые результаты из работ [5], [7], [8], [14], [19].

**А.** Функция $f(\Psi) = \max\limits_{v\in Q} \min\limits_{u\in P} (\Psi,\ \Phi(u,\ v) - l)$ непрерывна по переменному $\Psi \in S \subset R^\nu$. Отсюда и из компактности множества $D\ (z_0)$ следует, что существует вектор $\Psi_1(z_0) \in D(z_0)$ такой, что

$$N_1 \leqq \max_{v\in Q} \min_{u\in P} (\Psi_1(z_0),\ \Phi(u,\ v) - l)\ . \tag{8}$$

Очевидно, функция $(\Psi_1(z_0),\ \Phi(u,\ v) - l)$ при фиксированном $z_0 \in M_\mu$ непрерывна по совокупности $(u,\ v)$. Поэтому существует вектор $v_0(z_0) \in Q$ такой, что для всех $u \in P$ (см. [20])

$$N_1 \leqq (\Psi_1(z_0),\ \Phi\ (u,\ v_0(z_0)) - l)\,. \tag{9}$$

Рассуждая так же, как и выше, убеждаемся в существовании вектора $\Psi_2(z_0) \in D(z_0)$, для которого

$$N_2 \leqq \min_{u\in P} \max_{v\in Q} (\Psi_2(z_0),\ \Phi(u,\ v) - l)\,.$$

Ясно, что для любого $u \in P$

$$\min_{u\in P} \max_{v\in Q} (\Psi_2(z_0),\ \Phi(u,\ v) - l) \leqq \max_{u\in Q} (\Psi_2(z_0),\ \Phi(u,\ v) - l)\ .$$

Пусть $u(s)$, $0 \leqq s \leqq 1$, — произвольная измеримая функция со значениями из множества $P$. Рассмотрим следующее уравнение относительно $v \in Q$:

$$\max_{v\in Q} (\Psi_2(z_0),\ \Phi(u(s),\ v) - l) = (\Psi_2(z_0),\ \Phi(u(s),\ v) - l)\ . \tag{10}$$

В силу непрерывности функции $(\Psi_2(z_0),\ \Phi(u,\ v) - l)$ по совокупности $(u,\ v)$ (при фиксированном $z_0$) решение уравнения (10), очевидно, существует. Наименьшее в лексикографическом смысле решение уравнения (10) обозначим через $\boldsymbol{v}_0(s)$, $0 \leqq s \leqq 1$. Рассуждая так же, как и в [5], [8], можно доказать измеримость функции $\boldsymbol{v}_0(s)$, $0 \leqq s \leqq 1$. Отметим, что $\boldsymbol{v}_0(s) = \boldsymbol{v}(\boldsymbol{z}_0, \boldsymbol{u}(s))$.

**Б.** Возможны два случая: 1) $N_1 - N_3 > 0$; 2) $N_1 - N_3 \leqq 0$, $N_2 - N_3 > 0$. Положим $N = N_1$, $\Psi(\boldsymbol{z}_0) = \Psi_1(\boldsymbol{z}_0)$, $\bar{\boldsymbol{v}}(s) = \boldsymbol{v}_0(\boldsymbol{z}_0)$, $0 \leqq s \leqq 1$, в случае 1) и $N = N_2$, $\Psi(\boldsymbol{z}_0) = \Psi_2(\boldsymbol{z}_0)$, $\bar{\boldsymbol{v}}(s) = \boldsymbol{v}_0(s)$, $0 \leqq s \leqq 1$, в случае 2). Имеем (см. (9), (10))

$$N \leqq (\Psi(\boldsymbol{z}_0), \Phi(\boldsymbol{u}(s), \bar{\boldsymbol{v}}(s)) - \boldsymbol{l}) \tag{11}$$

для всех $s \in [0, 1]$.

Для решения $\boldsymbol{z}(t)$, $0 \leqq t \leqq 1$, уравнения $\dot{\boldsymbol{z}} = \boldsymbol{C}\boldsymbol{z} + f(\boldsymbol{u}(t), \bar{\boldsymbol{v}}(t))$, $\boldsymbol{z}(0) = \boldsymbol{z}_0 \in R^n \setminus M$, имеем (см.[5])

$$F_i \Pi_i \boldsymbol{z}(t) = F_i \Pi_i e^{t\boldsymbol{C}} \boldsymbol{z}_0 + t\boldsymbol{a}_{i0} + \ldots + \frac{t^{k_i-1}}{(k_i-1)!} \boldsymbol{a}_{i,k_i-2} + \frac{t^{k_i}}{k_i!} (\boldsymbol{l} + \boldsymbol{l}_i)$$

$$+ \int_0^t \frac{(t-s)^{k_i-1}}{(k_i-1)!} \{F_i \Pi_i \boldsymbol{C}^{k_i-1} [f(\boldsymbol{u}(s), \bar{\boldsymbol{v}}(s)) - \boldsymbol{l}_i - \boldsymbol{l}] + g_i(t, s)\} \, ds,$$

где

$$\boldsymbol{a}_{ij} = F_i \Pi_i \boldsymbol{C}^j f(P, Q), \quad |g_i(t, s)| \leqq d\,(t-s) \leqq dt, \quad 0 \leqq s \leqq t \leqq 1,$$

$d$ — константа. Очевидно, при $0 \leqq t \leqq 1$ имеет место разложение

$$F_i \Pi_i e^{t\boldsymbol{C}} \boldsymbol{z}_0 = F_i \Pi_i \boldsymbol{z}_0 + t F_i \Pi_i \boldsymbol{C} \boldsymbol{z}_0 + \ldots + \frac{t^{k_i}}{k_i!} F_i \Pi_i \boldsymbol{C}^{k_i} \boldsymbol{z}_0 + \frac{t^{k_i}}{k_i!} \varphi_i(t, \boldsymbol{z}_0),$$

где $|\varphi_i(t, \boldsymbol{z}_0)| \leqq d_1 t\,(1 + |\boldsymbol{z}_0|)$, $d_1$ — константа. Пусть

$$\vartheta = \min \left\{1, \frac{\bar{N}}{2d}, \frac{\bar{N}}{4d_1 (1 + |\boldsymbol{z}_0|)}\right\}, \tag{12}$$

где $\bar{N} = N_1 - N_3$, в случае, когда $N_1 - N_3 > 0$, и $\bar{N} = N_2 - N_3$ в случае, когда $N_1 - N_3 \leqq 0$, $N_2 - N_3 > 0$. Отметим, что $\bar{N} > 0$. Ясно, что $|g_i(t, s)| \leqq d\vartheta \leqq \bar{N}/2$ при всех $s, t \in [0, \vartheta]$, $s \leqq t$.

Из условий теоремы и вышесказанного нетрудно получить следующее неравенство

$$(\Psi(\boldsymbol{z}_0), F_j \Pi_i \boldsymbol{z}(t)) \geqq \frac{\bar{N}}{4k!} t^k, \tag{13}$$

где $t \in [0, \vartheta]$, $i = 1, \ldots, m$, $\boldsymbol{u}(t) \in P$ — произвольное измеримое управление, $\boldsymbol{v}(t) = \bar{\boldsymbol{v}}(t)$, $\boldsymbol{z}_0 \in M_\mu$, $k = \max\limits_i k_i$.

Очевидно, $\xi_i(t) \geqq d_2\,|\Pi_i \boldsymbol{z}(t)|$, $(F_i \Pi_i \boldsymbol{z}(t), \Psi(\boldsymbol{z}_0)) \leqq |F_i \Pi_i \boldsymbol{z}(t)| \leqq d_3\,|\Pi_i \boldsymbol{z}(t)|$ где, $d_2$, $d_3$ — константы, $d_3 > \max\limits_i \|F_i\|$, $\|F_i\|$ — операторная норма линейного отображения $F_i$. Поэтому для всех $t \in (0, \vartheta]$, $i = 1, \ldots, m$, (см. (13))

$$\xi_i(t) > \delta t^k, \tag{14}$$

где $\delta$ — константа, $\delta < \min\{1, \bar{N}/4d_2 k!\}$. Увеличив, если надо, константу $d_1$, можно считать, что

$$\vartheta = \frac{a}{1 + |\boldsymbol{z}_0|}, \tag{15}$$

где $a$ — константа, $a \leqq 1$.

Построение убегающего управления $\boldsymbol{v}(t)$ и получение оценки снизу (3) может быть произведено аналогично доказательству теоремы 2 из [8].

**Замечания 1:** В предположении 1 требуется непустота множества $D(\boldsymbol{z}_0)$, $\boldsymbol{z}_0 \in M_\mu$. Можно показать, что если $\sum_{i=1}^{m} k_i + m \leqq \nu$, то множество $D(\boldsymbol{z}_0)$ непусто для всех $\boldsymbol{z}_0 \in R^n$. Действительно, так как число неравенств, которым должны удовлетворять векторы $\Psi \in D(\boldsymbol{z}_0)$, равно $\sum_{i=1}^{m} k_i + m$, а размерность пространства $R^\nu$ удовлетворяет условию $\nu \geqq \sum_{i=1}^{m} k_i + m$, то множество $D(\boldsymbol{z}_0)$ непусто для всех $\boldsymbol{z}_0 \in R^n$ (см. [19]).

**В**. Рассуждая так же, как в п. **А**, можно убедиться в возможности построения измеримой функции $\bar{\boldsymbol{v}}_0(t)$, $0 \leqq t \leqq 2\vartheta$ (причем $\bar{\boldsymbol{v}}_0(t)$ – либо постоянный вектор $\boldsymbol{v}_0(\boldsymbol{z}_0)$, либо $\bar{\boldsymbol{v}}_0(t) = \boldsymbol{v}(\boldsymbol{z}_0, \boldsymbol{u}(t))$), для которой при всех $s \in [0, 2\vartheta]$

$$N \leqq (\Psi_0(\boldsymbol{z}_0), \Phi(\boldsymbol{u}(s), \bar{\boldsymbol{v}}_0(s)) - \boldsymbol{l}) , \tag{16}$$

где $N = N_1$ при $N_1 - N_3 > 0$ и $N = N_2$, если $N_1 - N_3 \leqq 0$, $N_2 - N_3 > 0$, $\Psi_0(\boldsymbol{z}_0)$ – некоторый вектор из $E(\boldsymbol{z}_0)$. Для решения $\boldsymbol{z}(t)$, $0 \leqq t \leqq 2\vartheta$, соответствующего функциям $\boldsymbol{u}(t)$, $\bar{\boldsymbol{v}}_0(t)$, $0 \leqq t \leqq 2\vartheta$, и начальному условию $\boldsymbol{z}(0) = \boldsymbol{z}_0 \notin M$, имеем (см. п. **Б**)

$$F_i \Pi_i \boldsymbol{z}(t) = G_i(\boldsymbol{z}_0, t) + \int_0^t \frac{(t-s)^{k_i - 1}}{(k_i - 1)!} \{[\Phi(\boldsymbol{u}(s), \bar{\boldsymbol{v}}_0(s)) - \boldsymbol{l}] + [\Phi_i(\boldsymbol{u}(s), \bar{\boldsymbol{v}}_0(s)) - \boldsymbol{l}_i] + g_i(t, s)\} \, \mathrm{d}s ,$$

где

$$G_i(\boldsymbol{z}_0, t) = F_i \Pi_i e^{t\boldsymbol{C}} \boldsymbol{z}_0 + t F_i \Pi_i f(P, Q) + \ldots + \frac{t^{k_i - 1}}{(k_i - 1)!} F_i \Pi_i \boldsymbol{C}^{k_i - 2} f(P, Q) + \frac{t^{k_i}}{k_i!} (\boldsymbol{l} + \boldsymbol{l}_i) ,$$

$|g_i(t, s)| \leqq d\,(t - s) \leqq dt$, $0 \leqq s \leqq t \leqq 2\vartheta$, $d$ – константа. Учитывая (16), аналогично п. **Б** получим

$$(F_i \Pi_i \boldsymbol{z}(t), \Psi_0(\boldsymbol{z}_0)) \geqq \frac{t^{k_i}}{2k_i!} \bar{N}, \quad 0 \leqq t \leqq 2\vartheta_1, \quad \vartheta_1 = \min\left\{\vartheta, \frac{\bar{N}}{4d}\right\},$$

где $\bar{N} = N_1 - N_3$ при $N_1 - N_3 > 0$ и $\bar{N} = N_2 - N_3$, если $N_1 - N_3 \leqq 0$, $N_2 - N_3 > 0$. Построение убегающего управления $\boldsymbol{v}(t)$ и получение оценки снизу (4) может быть произведено аналогично доказательству теоремы 6 из [8].

**Г**. Теорема 3 настоящей работы несколько усиливает результат теоремы 5 из [8]. Перейдем к доказательству теоремы 3.

Нетрудно видеть, что $\tilde{N}_1 \leqq \tilde{N}_2$. Пусть $\boldsymbol{z}_0 \in M_\mu$.

Функция

$$f(\Psi_1, \ldots, \Psi_m) = \max_{\boldsymbol{v} \in Q} \min_{\boldsymbol{u} \in P} \min_i (\Psi_i, \pi_i \boldsymbol{C}^{k_i - 1} f(\boldsymbol{u}, \boldsymbol{v}) - \boldsymbol{l}_i)$$

непрерывна по совокупности $\Psi_1, \ldots, \Psi_m$, принадлежащих $R^n$. Отсюда и из компактности $D_1(z_0), \ldots, D_m(z_0)$ следует, что существуют такие векторы $\varphi_1(z_0) \in D_1(z_0), \ldots, \varphi_m(z_0) \in D_m(z_0)$, что

$$\max_{\substack{\Psi_1 \in D_1(z_0) \\ \cdots\cdots \\ \Psi_m \in D_m(z_0)}} \max_{v \in Q} \min_{u \in P} \min_i \left(\Psi_i, \pi_i C^{k_i-1} f(u, v) - l_i\right)$$
$$= \max_{v \in Q} \min_{u \in P} \min_i \left(\varphi_i(z_0), \pi_i C^{k_i-1} f(u, v) - l_i\right).$$

Положим $\Psi_1^{j_1} = \varphi_1(z_0), \ldots, \Psi_m^{j_1} = \varphi_m(z_0)$. Функция $g(\Psi_1, \ldots, \Psi_m) = \min_{u \in P} \max_{v \in Q} \min_i$ $(\Psi_i, \pi_i C^{k_i-1} f(u, v) - l_i)$ непрерывна по совокупности $\Psi_1, \ldots, \Psi_m$, принадлежащих $R^n$. Отсюда и из компактности $D_1(z_0), \ldots, D_m(z_0)$ следует, что существуют такие векторы $\tilde{\varphi}_1(z_0) \in D_1(z_0), \ldots, \tilde{\varphi}_m(z_0) \in D_m(z_0)$, что

$$\max_{\substack{\Psi_1 \in D_1(z_0) \\ \cdots\cdots \\ \Psi_m \in D_m(z_0)}} \min_{u \in P} \max_{v \in Q} \min_i \left(\Psi_i, \pi_i C^{k_i-1} f(u, v) - l_i\right)$$
$$= \min_{u \in P} \max_{v \in Q} \min_i \left(\tilde{\varphi}_i(z_0), \pi_i C^{k_i-1} f(u, v) - l_i\right).$$

Положим $\Psi_1^{j_2} = \tilde{\varphi}_1(z_0), \ldots, \Psi_m^{j_2} = \tilde{\varphi}_m(z_0)$. Используя векторы $\Psi_1^{j_1}, \ldots, \Psi_m^{j_1}$, $\Psi_1^{j_2}, \ldots, \Psi_m^{j_2}$, можно далее почти дословно повторить рассуждения доказательства теоремы 5 из [8] для построения управления убегания и получения соответствующих оценок.

**Замечание 2:** Отметим, что из выполнения предположения 6 работы [8] следует выполнение предположения 3 настоящей работы. Для доказательства этого факта достаточно показать, что в случае выполнения предположения 6 числа $\tilde{N}_1, \tilde{N}_2$ из предположения 3 находятся в следующих отношениях с числами $N_1, N_2$ из предположения 6 ($W_i$, $L_i$ и $l_i \in W_i$ берутся одни и те же, $\tilde{W}_i$ — ортогональное дополнение $W_i$ в $L_i$, $\mu > 0$ — произвольное фиксированное число): $\tilde{N}_1 \geqq N_1$, $\tilde{N}_2 \geqq N_2$. Докажем неравенство $\tilde{N}_1 \geqq N_1$. Неравенство $\tilde{N}_2 \geqq N_2$ доказывается аналогично.

Фиксируем $z_0 \in R^n$. Из предположения 6 следует, что существуют такие векторы $\tilde{\varphi}_i(z_0) \in W_i$, $|\tilde{\varphi}_i(z_0)| = 1$, что

$$(\tilde{\varphi}_i(z_0), \pi_i z_0) = 0, \ldots, (\tilde{\varphi}_i(z_0), \pi_i C^{k_i-1} z_0 + \pi_i C^{k_i-2} f(P, Q)) = 0,$$
$$(\tilde{\varphi}_i(z_0), \pi_i C^{k_i} z_0 + l) = 0,$$
$$N_1 \leqq \max_{v \in Q} \min_{u \in P} \min_i \left(\tilde{\varphi}_i(z_0), \pi_i C^{k_i-1} f(u, v) - l_i\right).$$

Очевидно, $\tilde{\varphi}_i(z_0) \in D_i(z_0)$. Из сказанного и из определения $\tilde{N}_1$ легко следует, что $\tilde{N}_1 \geqq N_1$.

Теорема 4 доказывается аналогично теореме 3 с очевидными видоизменениями. При выводе оценки снизу для $\xi_i(t)$ следует воспользоваться рассуждениями § 6 работы Л. С. Понтрягина [1].

Теорема 5 настоящей работы несколько усиливает результат теоремы 6 из [8]. Перейдем к доказательству теоремы 5.

Нетрудно видеть, что $N_1^* \leqq N_2^*$. Фиксируем $z_0 \in R^n \backslash M$. Функция

$$f(\Psi_1, \ldots, \Psi_m) = \max_{v \in Q} \min_{u \in P} \min_i \left(\Psi_i, \pi_i C^{k_i - 1} f(u, v) - l_i\right)$$

непрерывна по совокупности $\Psi_1, \ldots, \Psi_m$, принадлежащих $R^n$. Отсюда и из компактности $E_1(z_0), \ldots, E_m(z_0)$ следует, что существуют такие векторы $\varphi_1(z_0) \in E_1(z_0), \ldots, \varphi_m(z) \in E_m(z_0)$, что

$$\max_{\substack{\Psi_1 \in E_1(z_0) \\ \cdots\cdots\cdots \\ \Psi_m \in E_m(z_0)}} \max_{v \in Q} \min_{u \in P} \min_i \left(\Psi_i, \pi_i C^{k_i - 1} f(u, v) - l_i\right)$$
$$= \max_{v \in Q} \min_{u \in P} \min_i \left(\varphi_i(z_0), \pi_i C^{k_i - 1} f(u, v) - l_i\right).$$

Функция $g(\Psi_1, \ldots, \Psi_m) = \min\limits_{u \in P} \max\limits_{v \in Q} \min\limits_i \left(\Psi_i, \pi_i C^{k_i - 1} f(u, v) - l_i\right)$ непрерывна по совокупности $\Psi_1, \ldots, \Psi_m$, принадлежащих $R^n$. Отсюда и из компактности $E_1(z_0), \ldots, E_m(z_0)$ следует, что существуют такие векторы $\tilde{\varphi}_1(z_0) \in E_1(z_0), \ldots, \tilde{\varphi}_m(z_0) \in E_m(z_0)$, что

$$\max_{\substack{\Psi_1 \in E_1(z_0) \\ \cdots\cdots\cdots \\ \Psi_m \in E_m(z_0)}} \min_{u \in P} \max_{v \in Q} \min_i \left(\Psi_i, \pi_i C^{k_i - 1} f(u, v) - l_i\right)$$
$$= \min_{u \in P} \max_{v \in Q} \min_i \left(\tilde{\varphi}_i(z_0), \pi_i C^{k_i - 1} f(u, v) - l_i\right).$$

Используя векторы $\varphi_1(z_0), \ldots, \varphi_m(z_0), \tilde{\varphi}_1(z_0), \ldots, \tilde{\varphi}_m(z_0)$, нетрудно осуществить построение векторов $\Psi_i^0 \in W_i$ единичной длины и специального управления $\bar{v}_0(t) \in Q$, $0 \leqq t \leqq 2\vartheta$ (причем $v_0(t)$ – либо постоянный вектор $v_0(z_0)$, либо $\bar{v}_0(t) = v(z_0, u(t))$), для которых при $s \in [0, 2\vartheta]$, $i = 1, \ldots, m$, выполняется неравенство (47) из [8]: $N \leqq \left(\Psi_i^0, \pi_i C^{k_i - 1} f(u(s), \bar{v}_0(s)) - l_i\right)$, где $N = N_1^*$, если $N_1^* > 0$, и $N = N_2^*$, если $N_1^* \leqq 0$, $N_2^* > 0$. Дальнейшие рассуждения можно провести по схеме доказательства теоремы 6 из [8].

**Замечание 3:** Отметим, что из выполнения предположения 7 работы [8] следует выполнение предположения 5 настоящей работы. Для доказательства этого факта достаточно показать, что в случае выполнения предположения 7 числа $N_1^*, N_2^*$ из предположения 5 находятся в следующих соотношениях с числами $N_1, N_2$ из предположения 7 ($W_i$, $L_i$ и $l_i \in W_i$ берутся одни и те же, $\tilde{W}_i$ – ортогональное дополнение $W_i$ в $L_i$): $N_1^* \geqq N_1$, $N_2^* \geqq N_2$. Доказательство последних неравенств производится аналогично доказательству неравенств $\tilde{N}_1 \geqq N_1$, $\tilde{N}_2 \geqq N_2$ в замечании 2 с очевидными видоизменениями.

## § 3. Примеры

**А**. Рассмотрим аналог обобщенного контрольного примера Л. С. Понтрягина (см. [1], [8]). Точки $\boldsymbol{x}_1, \ldots, \boldsymbol{x}_m$ преследуют точку $\boldsymbol{y}(\boldsymbol{x}_1, \ldots, \boldsymbol{x}_m, \boldsymbol{y} \in R^\nu)$. Преследование считается оконченным, если по крайней мере для одного значения индекса $i$ имеет место равенство $\boldsymbol{x}_i = \boldsymbol{y}$. Движения точек $\boldsymbol{x}_1, \ldots, \boldsymbol{x}_m$ и $\boldsymbol{y}$ описываются уравнениями

$$\boldsymbol{x}_i^{(p_i)} + a_{i1}\boldsymbol{x}_i^{(p_i-1)} + \ldots + a_{i,p_i}\boldsymbol{x}_i = \boldsymbol{u}_i, \quad \boldsymbol{u}_i \in P_i, \tag{17}$$

$$\boldsymbol{y}^{(q)} + b_1\boldsymbol{y}^{(q-1)} + \ldots + b_q\boldsymbol{y} = \boldsymbol{v}, \quad \boldsymbol{v} \in Q, \tag{18}$$

где $a_{ij}, b_k$ — линейные отображения $R^\nu$ в себя, $j = 1, \ldots, p_i$, $k = 1, \ldots, q$, $P_i$ и $Q$ — непустые компактные подмножества $R^\nu$.

Имеет место следующее утверждение (ср. с [8]).

**Утверждение:** *Пусть для уравнений* (17), (18): а) $(q+1)\,m \leqq \nu$; б) $q < p_j$ *для* $j = 1, \ldots, m_1$ *и существует вектор* $\boldsymbol{l} \in R^\nu$ *такой, что* $\min\limits_{\Psi \in S} \max\limits_{\boldsymbol{v} \in Q} (\Psi, \boldsymbol{v} - \boldsymbol{l}) > 0$; в) $q = p_j$ *для* $j = m_1 + 1, \ldots, m$ *и существуют векторы* $\boldsymbol{l}_j \in R^\nu$ *такие, что* $\min\limits_{\Psi \in S} \max\limits_{\boldsymbol{v} \in Q} (\Psi, \boldsymbol{v} - \boldsymbol{l}) > \max\limits_{\substack{\Psi \in S, \boldsymbol{u}_j \in P_j, \\ j = m_1 + 1, \ldots, m}} (\Psi, -\boldsymbol{u}_j - \boldsymbol{l}_j)$ (*случай* $m_1 = 0$, $m_1 = m$ *допускаются*). *Тогда для игры* (17), (18) *выполнены все условия теоремы* 1, *т.е. в игре* (17), (18) *возможно убегание.*

Справедливость утверждения доказывается путем непосредственных подсчетов, полагая $L_i$ ортогональным $M_i$, $W_i = L_i$, $\Phi(\boldsymbol{u}, \boldsymbol{v}) = \boldsymbol{v}$, $\Phi_i(\boldsymbol{u}, \boldsymbol{v}) = -\boldsymbol{u}_i$, где $i = m_1 + 1, \ldots, m$.

**Б**. Рассмотрим частный случай игры (17), (18) — аналог контрольного примера Л. С. Понтрягина (см. [1], [7], [8]). Уравнения (17), (18) имеют вид

$$\ddot{\boldsymbol{x}}_i + \alpha_i \dot{\boldsymbol{x}}_i = \varrho_i \overline{\boldsymbol{u}}_i, \quad |\overline{\boldsymbol{u}}_i| \leqq 1, \quad \ddot{\boldsymbol{y}} + \beta \dot{\boldsymbol{y}} = \sigma \overline{\boldsymbol{v}}, \quad |\overline{\boldsymbol{v}}| \leqq 1, \tag{19}$$

где $\alpha_i$, $\beta$, $\varrho_i$ и $\sigma$ — неотрицательные числа. Нетрудно показать, что все условия сформулированного выше утверждения выполнены, если (ср. с [8])

$$\nu \geqq 3m, \quad \sigma > \max_i \varrho_i. \tag{20}$$

Здесь $q = 2$, $m_1 = 0$.

**В**. Рассмотрим частный случай игры (17), (18) — аналог задачи о « мальчике и крокодиле » (см. [1], [7], [8]). Уравнения (17), (18) имеют вид

$$\ddot{\boldsymbol{x}}_i = \boldsymbol{u}_i, \quad |\boldsymbol{u}_i| \leqq 1, \quad \dot{\boldsymbol{y}} = \boldsymbol{v}, \quad |\boldsymbol{v}| \leqq 1. \tag{21}$$

Можно убедиться, что все условия утверждения выполнены, если (ср. с [8])

$$\nu \geqq 2m. \tag{22}$$

Здесь $q = 1$, $m_1 = m$. Отметим, что при выполнении условий (20), (22) для примеров (19), (21) справедлива и теорема 2. Далее, в случае, когда в (17), (18)

$a_{ij}, b_k$ — действительные числа, используя основные идеи доказательства теоремы 1, можно получить более тонкое, нежели приведенное в утверждении, достаточное условие для возможности убегания, связывающее коэффициенты $a_{ij}, b_k$, однако на этом мы здесь не будем останавливаться (см. [19]).

**Г.** Рассмотрим следующий простой пример

$$\dot{z}_1 = \boldsymbol{v}, \quad \dot{z}_2 = 0,$$

где $|\boldsymbol{v}| \leqq 1$, терминальное множество состоит из точки $0, z \in R^2$. Здесь $m = 1$, $P$ состоит из точки 0, $\boldsymbol{C} = 0$, $L_1 = R^2$. Положим $W_1 = L_1$, $\boldsymbol{l} = 0$, $\boldsymbol{l}_1 = 0$, $k_1 = 1$, $F_1 = E$, тогда $\Pi_1 = \pi_1 = \boldsymbol{E}$ — единичному оператору, $\pi_1 \boldsymbol{C}^{k_1-1} f(\boldsymbol{u}, \boldsymbol{v}) - \boldsymbol{l}_1 = \begin{pmatrix} \boldsymbol{v} \\ 0 \end{pmatrix}$, $D(z_0) = \boldsymbol{E}(z_0) = D_1(z_0) = \boldsymbol{E}_1(z_0)$ и они состоят из векторов $\Psi \in R^2$, $|\Psi| = 1$, таких, что $(\Psi, z_0) \geqq 0$. Положим $\Phi(\boldsymbol{u}, \boldsymbol{v}) = \begin{pmatrix} \boldsymbol{v} \\ 0 \end{pmatrix}$, $\Phi_1(\boldsymbol{u}, \boldsymbol{v}) = \begin{pmatrix} 0 \\ 0 \end{pmatrix}$. Нетрудно видеть, что для выбранных $L_1, W_1, \boldsymbol{l}, \boldsymbol{l}_1, F_1, \Phi(\boldsymbol{u}, \boldsymbol{v}), \Phi_1(\boldsymbol{u}, \boldsymbol{v})$

$$N_1 = N_2 = \tilde{N}_1 = \tilde{N}_2 = \tilde{\tilde{N}}_1 = \tilde{\tilde{N}}_2 = N_1^* = N_2^*, \quad N_3 = 0$$

и что $N_1 = \min\limits_{z_0, |z_0| = 1} \max\limits_{\Psi, |\Psi| = 1, (\Psi, z_0) \geqq 0} |\Psi_1|$, где $\Psi = \begin{pmatrix} \Psi_1 \\ \Psi_2 \end{pmatrix}$.

Можно показать, что $h(z_0) = \max\limits_{\Psi, |\Psi| = 1, (\Psi, z_0) \geqq 0} |\Psi_1| > 0$ $\forall z_0 \in R^2$ и зависит непрерывно от $z_0$ на окружности $|z_0| = 1$. Отсюда следует, что $N_1 > 0$ и, следовательно, выполнены предположения 1–5. Вместе с тем в этом примере не выполнены предположения 6, 7 из [8].

### Литература


[1] Понтрягин, Л. С.: Линейная дифференциальная игра убегания. Труды МИАН СССР им. В. А. Стеклова, т. **112**, 1971.

[2] Понтрягин, Л. С., и Е. Ф. Мищенко: Задача об уклонении от встречи в линейных дифференциальных играх. Дифференциальные уравнения **7** (1971) № 3.

[3] Pontryagin, L.: On the Evasion Process in Differential Games. Applied Mathematics and Optimization , **1** (1974) No. 1.

[4] Красовский, Н. Н.: К игровой задаче уклонения. Дифференциальные уравнения **8** (1972) № 2.

[5] Мищенко, Е. Ф., и Н. Сатимов: Задача об уклонении от встречи в дифференциальных играх с нелинейными управлениями. Дифференциальные уравнения **9** (1973) № 10.

[6] Gamkrelidze, R. V., G. L. Kharatishvili: A Differential Game of Evasion with Nonlinear Control. SIAM Journal on Control **12** (1974) No 2.

[7] Мищенко, Е. Ф., и Н. Сатимов: Задача уклонения от встречи в дифференциальных играх многих лиц. ДАН СССР **224** (1975) № 2.

[8] Мищенко, Е. Ф., М. С. Никольский и Н. Сатимов: Задача уклонения от встречи в дифференциальных играх многих лиц. Труды МИАН СССР им. В. А. Стеклова, т. **143**, **1977**.

[9] Пшеничный, Б. Н.: О задаче убегания. Кибернетика, (1975) № 4.

[10] Пшеничный, Б. Н.: Простое преследование несколькими объектами. Кибернетика, (1976) № 3.

[11] Черноусько, Ф. Л.: Одна задача уклонения от многих преследователей. Прикладная математика и механика **40** (1976) № 1.
[12] Лагунов, В. Н.: Нелинейная дифференциальная игра убегания. ДАН СССР **202** (1972) № 3.
[13] Чикрий, А. А.: Задача уклонения в нелинейных дифференциальных играх. Кибернетика (1975) № 3.
[14] Гусятников, П. Б.: Об $l$-уклонении от встречи в линейных дифференциальных играх. Прикладная математика и механика **38** (1974) № 3.
[15] Гусятников, П. Б.: $l$-убегание в линейных дифференциальных играх. Дифференциальные уравнения **12** (1976) № 3.
[16] Никольский, М. С.: О линейной задаче убегания. ДАН СССР **218** (1974) № 5.
[17] Никольский, М. С.: О квазилинейной задаче убегания. ДАН СССР **221** (1975) № 3.
[18] Сатимов, Н.: К задаче убегания в дифференциальных играх с нелинейными управлениями. ДАН СССР **216** (1974) № 4.
[19] Сатимов, Н.: Об одном способе уклонения от встречи в дифференциальных играх. Математический сборник **99** (141) (1976) выпуск 3.
[20] Демьянов, В. Ф., и В. Н. Малоземов: Введение в минимакс. « Мир », Москва, 1972.

**Zusammenfassung**

Es werden Lösungsmethoden für die Aufgabe über die Vermeidung des Zusammentreffens eines flüchtenden Objekts mit seinen Verfolgern entwickelt. Die Effektivität der erzielten Ergebnisse wird an einer Reihe von Beispielen allgemeinen Typs demonstriert. Weiterhin werden für die im Prozeß des Verfolgungsspiels auftretenden Abstände des flüchtenden Objekts zu jedem der einholenden Objekte untere Schranken angegeben.

**Summary**

Methods are constructed to solve the problem of evading the encounter between the controlled object and the group of pursuit objects. The efficiency of obtained results is demonstrated on some examples of general type. The lower estimates are obtained for distance between the evading object and every pursuit object in the process of evading from the encounter.

Received August 1979

## Book Review

G. F. Newell: **Approximate Behavior of Tandem Queues.** Lectures Notes in Economics and Mathematical Systems, Vol. 171. Springer-Verlag, Berlin-Heidelberg-New York 1979., 410 S., DM 39,–.

The monograph deals with the approximate stochastic behavior of a system consisting of a sequence of servers in series with finite storage between consecutive servers. More exactly, customers are considered to originate from an infinite reservoir. They first pass through a server with service rate $\mu_1$, variance rate $\Delta_1$, into a storage of capacity $c_1$, etc., until finally, after passing through an $n$th server, they disappear. Such so-called tandem queueing problems arise in specific transportation systems. (The monograph under discussion was originated by the problem of determining the effect of a finite block length on the overall service rate of a sequence of traffic signals along a highway.) Similar types of problems exist in the analysis of telephone switching networks.

The aim of the monograph is not to develop mathematical methods for their own sake. "The motivation was to develop techniques and describe phenomena relevant to the initial design of transportation systems, or, perhaps more important, to the prediction of the consequences of modifications in an existing system. ... To analyse a real system one does not wish to waste time to analysing parts of the system which cause no troubles. One wants to identify the critical points of the system; one wants simple tools for simple problems and to recognize when a problem is not simple." Thus, having in mind that the behavior of a multi-server system is usually dominated by the behavior of one or two critical servers (bottlenecks), the book deals essentially with the analysis of systems consisting of one or two servers. In this case one can principally get exact solutions of the diffusion equation which describes the behavior of the system. However, the exact solutions are quite difficult to derive and even more difficult to evaluate. Hence approximate solutions are presented and their properties are discussed. Chapter 1 deals with graphical representations and deterministic approximations. The diffusion equation being basic for the whole problem is introduced. A general solution of this partial differential equation is given. Chapter 2 is subjected to systems with a single server. The chapters 2–7 deal with systems consisting of two servers. Special cases are discussed, e.g. $\mu_1 = \mu_2$, $\Delta_2 = 0$ and $c_1 = c_2$. There are investigated among else (equilibrium) queue length distributions, service rates and variances (for the system) as well as marginal distributions. Never mind of the final conclusion of the author "that any hope of finding simple and general solutions to tandem queuing systems has been shattered", many hints for further analytic research are given.

The monograph contains a lot of new and interesting material. To study it some knowledge in probability theory and in analysis (solution of partial differential equations, Laplace transformation) is necessary. It can be recommended to both the theoretician and the practician.

F. Beichelt

Math. Operationsforsch. Statist., Ser. Optimization, Vol. 11 (1980) No. 4, 593–603

# Lösung strukturierter Aufgaben der optimalen Steuerung mittels Zerlegungsverfahren

B. LUDERER [1]

**Zusammenfassung:** Für ein konvexes Optimierungsproblem in HILBERTräumen wird eine Lösungsmethode beschrieben, die auf dem Prinzip der Zielfunktionssteuerung beruht. Dieser allgemeine Zugang wird danach auf ein linear-quadratisches Optimalsteuerproblem angewendet, das eine bestimmte Struktur besitzt. Dabei entstehen Teilaufgaben mit geringerer Dimension und ohne Zustandsgleichungen, die wiederholt gelöst werden müssen. Abschließend werden einige numerische Erfahrungen mitgeteilt.

## 1. Einleitung

In der vorliegenden Arbeit wird für Probleme der optimalen Steuerung, die eine bestimmte Struktur besitzen, ein Lösungsverfahren vorgestellt, das auf einem dualen Zugang beruht. Dies bietet den Vorteil, daß die im Laufe des Rechenprozesses zu lösenden Teilaufgaben eine einfachere Gestalt aufweisen und bei entsprechender Struktur in eine Reihe kleinerer Teilaufgaben zerfallen. Das Verfahren gehört also zu den Dekompositionsmethoden, die z. B. in [1] für Aufgaben in endlichdimensionalen Räumen sehr ausführlich dargestellt sind. Die vorliegenden Untersuchungen stellen in gewisser Hinsicht eine Übertragung dort beschriebener Ideen auf Probleme der optimalen Steuerung dar. Ein derartiger Zugang wird auch in [8] beschrieben. Andere Dekompositionsmöglichkeiten für ähnliche Aufgabenstellungen werden in [2] (unter Benutzung des Wechselwirkungs-Voraussage-Prinzips) und [6] (unter Ausnutzung notwendiger Optimalitätsbedingungen und eines Fixpunktalgorithmus) vorgeschlagen; ein direkter Vergleich ist aber auf Grund unvollständig veröffentlichter bzw. fehlender numerischer Resultate nicht möglich. Weitere Möglichkeiten der Anwendung von Verfahren der mathematischen Optimierung auf Optimalsteuerprobleme kann man z. B. in [11] und [7] finden.

[1] Technische Hochschule Karl-Marx-Stadt, Sektion Mathematik, DDR - 9022 Karl-Marx-Stadt, Reichenhainer Str. 41.

## 2. Problemstellung in Hilberträumen

Für die konvexe Optimierungsaufgabe

$$\begin{cases} f(z) \to \inf; \\ F(z) = 0 , \\ z \in Z_0 , \end{cases} \tag{1}$$

wobei $f: Z \to R$ ein konvexes stetiges Funktional, $F: Z \to Y$ eine stetige lineare Abbildung, $Y$ und $Z$ Hilberträume und $Z_0$ eine konvexe Teilmenge von $Z$ sein sollen, läßt sich mit Hilfe der zugehörigen Lagrange-*Funktion*

$$\mathfrak{L}(z, y^*, \lambda_0) = \lambda_0 f(z) + \langle y^*, F(z) \rangle, \quad \lambda_0 \in R, \quad y^* \in Y^* \tag{2}$$

und durch Einführung der Funktion

$$\varphi(y^*) = \inf_{z \in Z_0} \mathfrak{L}(z, y^*, \lambda_0) \tag{3}$$

die zu (1) *duale Aufgabe* formulieren:

$$\varphi(y^*) \to \sup; \quad y^* \in Y^*. \tag{4}$$

Es soll vorausgesetzt werden, daß der optimale Zielfunktionswert von (1) endlich ist. Da $Y$ ein Hilbertraum ist, kann der Dualraum $Y^*$ mit $Y$ identifiziert werden. Offensichtlich ist $\varphi$ konkav. Wir schlagen vor, das Problem (4) mit Hilfe eines Subgradientenverfahrens zu lösen, welches auf Grund der fest vorgegebenen Schrittweite sehr einfach zu realisieren ist, während für die Bestimmung einer Lösung von (3) ein beliebiges Verfahren angewendet werden kann (im Laufe des Lösungsprozesses von (4) müssen mehrfach Aufgaben vom Typ (3) gelöst werden).

Selbstverständlich ist das Lösen der Dualaufgabe (4) anstelle der Ausgangsaufgabe (1) nur dann sinnvoll, wenn zwischen beiden Problemen in gewissem Sinne Äquivalenz besteht (Übereinstimmung der optimalen Zielfunktionswerte, Gewinnung einer optimalen Lösung von (1) aus einer optimalen Lösung von (4)). Bedingungen für das Übereinstimmen der optimalen Zielfunktionswerte und eine Beschreibung der dualen Lösungen findet man z. B. in [5] Kap. 7.3 (für den Fall $Y = R^m$); dort werden auch Bedingungen angegeben, unter denen der Lagrange*sche Multiplikator* $\lambda_0$ von Null verschieden ist (*regulärer Fall*). Es ist dann sinnvoll, Aufgabe (4) zu lösen, wenn dies irgendeinen Vorteil mit sich bringt.

Wie oben erwähnt, wollen wir zur Maximierung von $\varphi$ ein Subgradientenverfahren einsetzen, welches ja zumindest prinzipiell im Hilbertraum anwendbar ist. Bezüglich der Beschreibung eines solchen Verfahrens sei auf [1], [9] verwiesen. Zu seiner Anwendung müssen wir in der Lage sein, einen Subgradienten explizit anzugeben, ohne jedoch das gesamte Subdifferential zu beschreiben. Im Falle der Existenz eines Minimums in der Aufgabe

$$\mathfrak{L}(z, y^*, 1) = f(z) + \langle y^*, F(z) \rangle \to \inf \tag{5}$$

ist dies leicht möglich. Falls das Infimum in (5) nicht angenommen wird, ist unklar, ob $\varphi$ im entsprechenden Punkt subdifferenzierbar ist bzw. wie ein Subgradient ermittelt werden kann. Wie im konkreten Fall diese Schwierigkeit umgangen werden kann, wird in Abschnitt 3 gezeigt.

**Satz 1:** *Es sei* $Z^*(y^*) = \{z \in Z_0 \mid f(z) + \langle y^*, F(z)\rangle = \varphi(y^*) = \min_{w \in Z_0} [f(w) + \langle y^*, F(w)\rangle]\}$. *Dann ist* $\hat{y} = F(\hat{z}) \in Y$ *ein Element des Subdifferentials der nach* (3) *definierten Funktion* $\varphi$ *im Punkt* $\hat{y}^*$, *wenn* $\hat{z} \in Z^*(\hat{y}^*)$.

Der Beweis beruht auf der Definition des Subdifferentials. ∎

**Satz 2:** *Es sei* $f(s, x)$ *eine auf* $S \times X$ *definierte Funktion, die konkav in* $x$ *für alle* $s \in S$ *ist, ferner sei* $s \to f(s, x)$ *unterhalbstetig für beliebiges* $x \in X$ *und* $x \to f(s, x)$ *stetig in* $x_0$ *für alle* $s \in S$. *Wir setzen* $\psi(x) = \inf_{s \in S} f(s, x)$, $S_0(x) = \{s \in S \mid f(s, x) = \psi(x)\}$. *Schließlich soll* dom $\psi = \{x \in X \mid \psi(x) > -\infty\} = X$ *sein und eine Umgebung* $U(x_0)$ *des Punktes* $x_0$ *sowie eine kompakte Menge* $K \subset S$ *existieren, so daß für alle* $x \in U(x_0)$ *die Menge* $S_0(x)$ *nichtleer und in* $K$ *enthalten ist. Dann gilt*

$$\overline{\operatorname{conv}}\,\Big(\bigcup_{s \in S_0(x_0)} \partial f_s(x_0)\Big) = \partial \psi(x_0)\ ,$$

*wobei* $f_s(x) = f(s, x)$ *auf* $X$ *definierte Funktionen sind.*

Beweis: Der Satz stellt die Modifikation eines in [5] Kap. 4.2 angegebenen Resultates dar. Der dort angegebene Beweis bleibt unter den geänderten Voraussetzungen im wesentlichen erhalten. ∎

**Satz 3:** *Die Optimallösungsmenge* $Z^*(\hat{y}^*) \subset Z$ *aus Satz 1 sei einelementig, und es existiere eine kompakte Menge* $K \subset Z$ *derart, daß* $Z^*(y^*) \subset K$ *für alle* $y^*$ *aus einer gewissen Umgebung von* $\hat{y}^*$. *Ferner sei die nach* (3) *definierte Funktion* $\varphi$ *stetig in* $\hat{y}^*$ *und* dom $\varphi = Y^*$. *Dann ist* $\varphi$ *in* $\hat{y}^*$ GÂTEAUX-*differenzierbar.*

Beweis: Auf Grund der Voraussetzungen des Satzes und der an Problem (1) gestellten Forderungen sind alle Bedingungen von Satz 2 mit $f(s, x) = \mathfrak{L}(z, y^*, \lambda_0)$ bzw. $\psi(x) = \varphi(y^*)$ erfüllt. Da $Z^*(\hat{y}^*) = \{\bar{z}\}$ einelementig ist, gilt $\partial\varphi(\hat{y}^*) = F(\bar{z})$, wie unschwer zu sehen ist. Gemäß entsprechender Aussagen der konvexen Analysis ist $\varphi$ somit in $\hat{y}^*$ GÂTEAUX-differenzierbar (vgl. [5] Kap. 4.2). ∎

## 3. Lösungsverfahren für ein linear-quadratisches Optimalsteuerproblem

Nunmehr kommen wir zu einer Konkretisierung der bisher von uns betrachteten allgemeinen Optimierungsaufgabe (1). Dabei handelt es sich um die folgende *Aufgabe der optimalen Steuerung mit linearer Zustandsgleichung und quadratischem Zielfunktional:*

$$\left\{\begin{aligned} &\mathfrak{J}(x(\cdot), u(\cdot)) = \int_{t_0}^{t_1} [x^{\mathsf T}(t)\,A(t)\,x(t) + u^{\mathsf T}(t)\,B(t)\,u(t)]\,\mathrm{d}t \to \inf\,;\\ &\dot{x}(t) = C(t)\,x(t) + D(t)\,u(t)\,;\\ &x(t_0) = x_0\,,\\ &u(t) \in U_0 \quad \forall t \in [t_0, t_1]\,. \end{aligned}\right. \tag{6}$$

Dabei seien die *Zustandsfunktionen* $x(t)$ Elemente des Raumes $W^n_{2,1}([t_0, t_1])$ der auf dem Intervall $[t_0, t_1]$ absolut stetigen Vektorfunktionen $x(t) = (x_1(t), \ldots, x_n(t))^{\mathsf{T}}$, deren erste Ableitungen dem Raum $L^n_2([t_0, t_1])$ angehören. Weiter seien die *Steuerungen* $u(t)$ aus $L^r_2([t_0, t_1])$ sowie $A(t)$, $B(t)$, $C(t)$, $D(t)$ stetig von $t$ abhängige Matrizen der Dimensionen $n \times n$, $r \times r$, $n \times n$, $n \times r$. Die Matrizen $A$ und $B$ seien symmetrisch sowie $A(t) > 0$, $B(t) > 0$ $\forall t \in [t_0, t_1]$. Schließlich mögen $U_0$ eine konvexe Teilmenge des $R^r$, $x_0$ ein gegebener Vektor aus dem $R^n$ und $t_0$, $t_1$ fixierte Zeitpunkte sein.

Aufgabe (6) läßt sich in der Form von Problem (1) darstellen, wenn wir folgende Bezeichnungen einführen:

$$X = W^n_{2,1}([t_0, t_1]), \quad U = L^r_2([t_0, t_1]), \quad Z = X \times U, \quad Y = L^n_2([t_0, t_1]),$$
$$X_0 = \{x(t) \in X \mid x(t_0) = x_0\}, \quad \mathfrak{U}_0 = \{u(t) \in U \mid u(t) \in U_0 \ \forall t \in [t_0, t_1]\},$$
$$Z_0 = X_0 \times \mathfrak{U}_0, \quad z = (x, u) = (x(\cdot), u(\cdot)),$$
$$F(z) = \dot{x}(t) - C(t)\, x(t) - D(t)\, u(t), \quad F: Z \to Y,$$
$$f(z) = \int_{t_0}^{t_1} [x^{\mathsf{T}}(t)\, A(t)\, x(t) + u^{\mathsf{T}}(t)\, B(t)\, u(t)]\, dt, \quad f: Z \to R.$$

**Bemerkung.** Wie sich leicht nachprüfen läßt, liegt für die Aufgabe (6) der reguläre Fall vor, d. h., man kann o. B. d. A. $\lambda_0 = 1$ setzen.

Das der Aufgabe (5) entsprechende spezielle Problem lautet:

$$\left\{\begin{aligned} &\mathfrak{L}(x(\cdot), u(\cdot), p(\cdot)) = \int_{t_0}^{t_1} [x^{\mathsf{T}}(t)\, A(t)\, x(t) + u^{\mathsf{T}}(t)\, B(t)\, u(t) + \\ &\qquad + p^{\mathsf{T}}(t)\, (\dot{x}(t) - C(t)\, x(t) - D(t)\, u(t))]\, dt \to \inf; \\ &x(t_0) = x_0, \\ &u(t) \in U_0 \ \forall t \in [t_0, t_1]. \end{aligned}\right. \tag{7}$$

Hierbei ist der LAGRANGEsche Multiplikator $p(t)$ ein Element des Raumes $L^n_2([t_0, t_1]) = Y^* = Y$. Die zu (6) duale Aufgabe ist dann

$$\varphi(p(\cdot)) \to \sup; \quad p(\cdot) \in L^n_2([t_0, t_1]), \tag{8}$$

wobei $\varphi(p(\cdot)) = \inf\limits_{(x(\cdot), u(\cdot)) \in X_0 \times \mathfrak{U}_0} \mathfrak{L}(x(\cdot), u(\cdot), p(\cdot))$ gilt.

### Prinzipieller Lösungsalgorithmus für (8)

1. Wähle Zahlen $\gamma_k > 0$, $k = 0, 1, \ldots$, $\sum\limits_{k=0}^{\infty} \gamma_k = \infty$, $\gamma_k \xrightarrow[k \to \infty]{} 0$ sowie eine Ausgangsfunktion $p_0(t)$. Setze $k := 0$.
2. Löse Aufgabe (7) mit vorgegebenem $p(t) = p_k(t)$ nach einem beliebigen Verfahren. Die optimale Lösung sei $(x_k(t), u_k(t))$. Berechne den „Defekt der Nebenbedingung" (Subgradient von $\varphi$ gemäß der Sätze 1 und 2):

   $$a_k(t) = \dot{x}_k(t) - C(t)\, x_k(t) - D(t)\, u_k(t)\,^2.$$

   Ist $a_k(t) \equiv 0$, so gehe zu 4., sonst zu 3.

---

[2] Wegen $x(\cdot) \in W^n_{2,1}$ existiert $\dot{x}(\cdot)$ nur fast überall. Folglich kann auch $a_k(\cdot)$ nur f. ü. berechnet werden.

3. Berechne $p_{k+1}(t) := p_k(t) + \gamma_k a_k(t)$ und kehre mit $k := k+1$ zurück nach 2.
4. Ende: $p_k(t)$ ist optimal für (8).

**Bemerkungen:** 1. Leider besitzt Aufgabe (7) nicht für beliebiges $p(\cdot) \in L_2^n$ ein endliches Infimum, so daß sich keine Aussage der Art dom $\varphi = Y^* = L_2^n$ treffen läßt. Dies zeigt das Beispiel

$$\mathfrak{I}(x(\cdot)) = \int_0^1 [x^2(t) + \dot{x}(t)]\, dt \to \inf; \quad x(0) = 0 .$$

Für die Folge $\{x_n(\cdot)\} \subset W_{2,1}([0, 1])$ mit

$$\dot{x}_n(t) = \begin{cases} 0, & 0 \leqq t \leqq (n-1)/n, \\ -n^2, & (n-1)/n \leqq t \leqq 1, \end{cases} \quad x_n(0) = 0$$

strebt $\mathfrak{I}(x_n(\cdot))$ mit wachsendem $n$ gegen $-\infty$. Um deshalb das Subgradientenverfahren ohne Bedenken anwenden zu können, müssen wir die Funktionenklasse $W_{2,1}^n$ verlassen. Wir können die Bedingung dom $\varphi = Y^*$ und damit die Anwendbarkeit des Algorithmus sichern, wenn wir beispielsweise mit einem Ritz-Ansatz arbeiten oder eine Diskretisierung der Zeit vornehmen (die Lösung also in gewissen Teilklassen von $W_{2,1}^n$ suchen).

2. Bei der Auswahl der Ausgangsfunktion $p_0(t)$ kann man gewisse qualitative Eigenschaften der optimalen Lösung oder andere Informationen — sofern sie bekannt sind — berücksichtigen, insbesondere sollte die *Transversalitätsbedingung* $p(t_1) = 0$ beachtet werden.

Wie läßt sich nun aus der Kenntnis einer optimalen Lösung von (8) eine optimale Lösung für (6) gewinnen? Um den Dualitätssatz aus [5] anwenden zu können, muß $Y = R^m$ sein (in diesem Fall läßt sich auch dom $\varphi = Y^*$ garantieren). Ist dann $p_*(t)$ eine mit obigem Algorithmus ermittelte optimale Lösung von (8) und $(x_*(t), u_*(t))$ optimale Lösung der zugehörigen Aufgabe (7), so gilt offenbar $a_*(t) \equiv 0$, und $(x_*(t), u_*(t))$ ist auch optimale Lösung von (6). Wurde jedoch $p_*(t)$ nicht exakt bestimmt, so kann $a_*(t)$ erheblich von Null abweichen, so daß $(x_*(t), u_*(t))$ nicht zulässig in (6) ist. Ein praktikabler Weg zur Umgehung dieser Schwierigkeit besteht in der Diskretisierung der Teilaufgaben (7).

**Bemerkung:** Auf Fragen der Stabilität des Problems im Sinne einer immer besseren approximativen Beschreibung aller vorkommenden Funktionen und Werte bei Verfeinerung der Diskretisierung soll im Rahmen dieser Arbeit nicht eingegangen werden; eine solche Stabilität wird jedoch im weiteren vorausgesetzt. Dieser Aspekt wird z. B. in [12] näher untersucht.

Das Intervall $[t_0, t_1]$ werde in $T$ gleiche Teilintervalle der Länge $h = (t_1 - t_0)/T$ zerlegt, wodurch die Diskretisierungspunkte $t^0 = t_0, t^1, \ldots, t^T = t_1$ entstehen. Die zugehörigen Werte der Steuer- und Zustandsvariablen bezeichnen wir mit $x^i = x(t^i)$ bzw. $u^i = u(t^i)$, $i = 1, \ldots, T$. Ferner seien $\mathcal{X} \subset R^{nT}$ und $\mathcal{V} \subset R^{rT}$ die

Mengen derjenigen $x=\{x_j^i\}_{i=1,j=1}^{T\ \ n}$ bzw. $u=\{u_j^i\}_{i=1,j=1}^{T\ \ r}$, die bei der eben beschriebenen Zuordnung von $x(t)\in W_{2,1}^n([t_0, t_1])$ bzw. $u(t)\in L_2^r([t_0, t_1])$ erzeugt werden. Für die Ableitung $\dot{x}(t)$ verwenden wir als Approximation den einfachstmöglichen Differenzenquotienten $\dot{x}^i=\dot{x}(t^i)=(x^i-x^{i-1})/h$. Analog betrachten wir anstelle des LAGRANGEschen Multiplikators $p(t)\in L_2^n([t_0, t_1])$ das zugeordnete $p=\{p_j^i\}_{i=1,j=1}^{T\ \ n}$. Die Menge aller solcher $p$ sei $\mathcal{P}\subset R^{nT}$. Entsprechend gehen auch die Matrizen $A(t)$, $B(t)$, $C(t)$, $D(t)$ in $A^i$, $B^i$, $C^i$, $D^i$, $i=1, \ldots, T$, über.

**Satz 4:** *Wird auf Aufgabe* (6) *bzw.* (7) *die oben beschriebene Diskretisierung angewendet und dabei h so gewählt, daß* det $(I-hC^i)\neq 0$, $i=1, \ldots, T$ *(I – Einheitsmatrix), so sind alle Bedingungen des Dualitätssatzes aus* [5] *Kap.* 7.3 *erfüllt.*

Beweis: Die Diskretisierung zieht die Endlichdimensionalität von $Y$ nach sich, während die Forderung det $(I-hC^i)\neq 0$ die Lösbarkeit des aus der Zustandsgleichung entstehenden linearen Gleichungssystems für beliebige rechte Seiten sichert, also die Regularität von $F$. Alle anderen Forderungen des Dualitätssatzes sind auf Grund der Voraussetzungen an (6) erfüllt. ∎

Nunmehr können wir Aufgabe (7) durch ein sie approximierendes endlichdimensionales Problem ($7_A$) ersetzen, wenn wir in (7) das Integralfunktional durch eine endliche Summe annähern (z. B. mit Hilfe der Rechteck- oder Trapezregel). Analog geht (8) in ($8_A$) über.

**Satz 5:** *In Aufgabe* ($7_A$) *existiert für beliebiges* $\bar{p}\in\mathcal{P}$ *eine eindeutige minimale Lösung.*

Beweis: Unter Berücksichtigung der Eigenschaft der positiven Definitheit von $A(t)$ und $B(t)$ (was sich auf $A^i$ bzw. $B^i$ in ($7_A$) überträgt und auch bei Verwendung entsprechender Formeln zur numerischen Integration erhalten bleibt) sowie der durch die Diskretisierung entstehenden Endlichdimensionalität läßt sich zeigen, daß der minimale Zielfunktionswert in ($7_A$) endlich ist und das Minimum eindeutig angenommen wird (vgl. [4]). ∎

Wir wollen zur Lösung von (8) eine Apriori-Diskretisierung in ($7_A$) benutzen; die entsprechenden Klassen werden mit $\mathcal{X}_0\times\mathcal{V}_0$ (bzw. $\mathcal{P}_0$) bezeichnet.

**Satz 6:** *Bei Benutzung einer festen Diskretisierung in* ($7_A$) *ist die über* $\mathcal{P}_0$ *definierte und* $\varphi$ *approximierende Funktion* $\tilde{\varphi}$ *stetig differenzierbar.*

Beweis: Für gegebenes $\bar{p}\in\mathcal{P}_0$ liegen die Lösungen von ($7_A$) innerhalb einer kompakten Menge $K(\bar{p})$ (vgl. [4]). Auf Grund der Endlichdimensionalität gilt ferner $K(p)\subset K$ für eine gewisse kompakte Menge $K$, wenn nur $p$ in einer Umgebung $U(\bar{p})$ liegt. Damit ist die entsprechende Voraussetzung aus Satz 3 für das vorliegende Problem erfüllt. Gemäß Satz 5 ist in $\mathcal{X}_0\times\mathcal{V}_0$ die Optimallösungsmenge nichtleer und einelementig; weiterhin gilt dom $\tilde{\varphi}=\mathcal{P}_0$. Wegen $Y=R^{nT}$ ist folglich $\tilde{\varphi}$ stetig über $\mathcal{P}_0$ (s. [5] Kap. 3.5), somit gemäß Satz 3 auch GÂTEAUX-differenzierbar und mithin sogar stetig differenzierbar (s. [10] Kap. 5). ∎

**Folgerung:** *Der Algorithmus erzeugt eine „fast optimale" Lösung in der Klasse* $\mathcal{X}_0 \times \mathcal{V}_0$.

Beweis: Ist $p_* \in \mathcal{P}_0$ eine optimale Lösung von $(8_A)$, so geht die für konvexe Funktionen gültige Optimalitätsbedingung $0 \in \partial\tilde{\varphi}(p_*)$ wegen der Behauptung des Satzes in $\tilde{\varphi}'(p_*)=0$ über. Da im allgemeinen aber $p_*$ nicht exakt bestimmt werden kann, gilt für die zu $p_*$ gehörige optimale Lösung von $(7_A)$ $z_*=(x_*, u_*) \in \in \mathcal{X}_0 \times \mathcal{V}_0$ die Beziehung $a_*=F(z_*) \neq 0$, d. h., $z_*$ ist nicht zulässig in $(6_A)$. Auf Grund der stetigen Differenzierbarkeit von $\tilde{\varphi}$ ist dann aber $\|F(z_*)\|_{R^{nT}}$ klein, so daß also $z_*$ „fast zulässig" im Sinne der Norm $\|\cdot\|_{R^{nT}}$ in der Klasse $\mathcal{X}_0 \times \mathcal{V}_0$ ist. ∎

**Bemerkungen.** 1. Wählen wir die Diskretisierung hinreichend fein und ist das Problem – wie vorausgesetzt – stabil, so erzeugen wir also mit dem Algorithmus auch eine fast optimale Lösung der Ausgangsaufgabe (6).

2. Im soeben beschriebenen Fall der Diskretisierung der Teilaufgaben stellt der vorgeschlagene Algorithmus kein Subgradientenverfahren im echten Sinne dar, da das Subdifferential jeweils nur ein Element enthält. Trotzdem ist das Verfahren natürlich anwendbar, umsomehr, als es wegen der a priori gegebenen Schrittweite relativ unkompliziert ist. Aber auch andere Verfahren (wie z. B. Gradientenmethoden) können eingesetzt werden.

3. Im allgemeinen Fall der Nichtdifferenzierbarkeit von $\varphi$ besteht eine Möglichkeit zur Bestimmung einer optimalen Lösung für (6) im Lösen einer Hilfsaufgabe, in der das normkleinste Element des Subdifferentials von $\varphi$ gesucht wird.

## 4. Dekomposition bei Vorhandensein einer Struktur

Nunmehr wollen wir untersuchen, wie sich das Vorhandensein einer bestimmten Struktur in der Ausgangsaufgabe (6) vorteilhaft auf die numerische Lösung auswirkt. Im Grunde genommen war gerade dieser Aspekt die Triebfeder für den vorgestellten Algorithmus.

Wir nehmen im weiteren an, daß für die Räume $X$ und $U$ die Darstellung $X=X_1\times\cdots\times X_N$, $U=U_1\times\cdots\times U_M$ gilt, d. h. $x(\cdot)=(x_1(\cdot), \ldots, x_N(\cdot))^{\mathsf{T}}$, $u(\cdot)= =(u_1(\cdot), \ldots, u_M(\cdot))^{\mathsf{T}}$. Die Menge $U_0$ lasse sich als $U_0^1\times\cdots\times U_0^M$ darstellen, und $C_{ij}$ bzw. $D_{ij}$ seien die der Zerlegung von $x(\cdot)$ bzw. $u(\cdot)$ entsprechenden Teilmatrizen von $C$ bzw. $D$. Weiterhin mögen $A(t)$ und $B(t)$ Blockdiagonalmatrizen sein, d. h.

$$A(t)=\begin{pmatrix} A_1(t) & & & \\ & A_2(t) & & \\ & & \ddots & \\ & & & A_N(t)\end{pmatrix}, \quad B(t)=\begin{pmatrix} B_1(t) & & \\ & \ddots & \\ & & B_M(t)\end{pmatrix}.$$

Dabei seien $A_i(t)>0$, $i=1, \ldots, N$, $B_j(t)>0$, $j=1, \ldots, M$, $\forall t\in[t_0, t_1]$. Entsprechend werde der LAGRANGEsche Multiplikator in $p(\cdot)=(p_1(\cdot), \ldots, p_N(\cdot))^{\mathsf{T}}$ aufgeteilt. Die zu (6) gehörende LAGRANGE-Funktion lautet dann:

$$\begin{aligned}\mathfrak{L}(x(\cdot), u(\cdot), p(\cdot)) &= \int_{t_0}^{t_1} [x^{\mathsf T}(t)\, A(t)\, x(t) + u^{\mathsf T}(t)\, B(t)\, u(t) \\ &\quad + p^{\mathsf T}(t)\, (\dot x(t) - C(t)\, x(t) - D(t)\, u(t))]\, \mathrm{d}t \\ &= \int_{t_0}^{t_1} \left\{ \sum_{i=1}^{N} \left[ x_i^{\mathsf T}(t)\, A_i(t)\, x_i(t) + p_i^{\mathsf T}(t) \left( \dot x_i(t) - \sum_{j=1}^{N} C_{ij}(t)\, x_j(t) \right) \right] \right. \\ &\quad \left. + \sum_{i=1}^{M} u_i^{\mathsf T}(t) B_i(t)\, u_i(t) - \sum_{i=1}^{N} p_i^{\mathsf T}(t) \sum_{j=1}^{M} D_{ij}(t)\, u_j(t) \right\} \mathrm{d}t \\ &= \sum_{i=1}^{N} \mathfrak{L}_{1i} + \sum_{i=1}^{M} \mathfrak{L}_{2i}\,,\end{aligned} \tag{9}$$

wobei

$$\mathfrak{L}_{1i} = \int_{t_0}^{t_1} (x_i^{\mathsf T}(t)\, A_i(t)\, x_i(t) + p_i^{\mathsf T}(t)\, \dot x_i(t) - q_i^{\mathsf T}(t)\, x_i(t))\, \mathrm{d}t\,,$$

$$\mathfrak{L}_{2i} = \int_{t_0}^{t_1} (u_i^{\mathsf T}(t)\, B_i(t)\, u_i(t) - w_i^{\mathsf T}(t)\, u_i(t))\, \mathrm{d}t$$

mit

$$q_i^{\mathsf T}(t) = \sum_{j=1}^{N} p_j^{\mathsf T}(t)\, C_{ji}(t), \quad w_i^{\mathsf T}(t) = \sum_{j=1}^{N} p_j^{\mathsf T}(t)\, D_{ji}(t)$$

gilt, d. h., $\mathfrak{L}$ zerfällt in eine Summe von *Teil-LAGRANGE-Funktionen.*

Das Lösen von (8) erfordert im $k$-ten Schritt die Minimierung der LAGRANGE-Funktion (9) bei vorgegebenem LAGRANGEschen Multiplikator $p_k(\cdot)$ unter denjenigen Nebenbedingungen, die nicht in die LAGRANGE-Funktion aufgenommen wurden. Die Aufgabe zerfällt in eine Reihe kleinerer Teilaufgaben. Der Lösungsprozeß vereinfacht sich mithin nicht nur dadurch, daß keine Nebenbedingungen der Form $\dot x = Cx + Du$ mehr vorkommen, sondern auch durch die Reduzierung der Dimension der entsprechenden Teilaufgaben. Im Schritt 2 des angegebenen Algorithmus sind folglich nachstehende Probleme zu lösen ($i=1, \ldots, N, j = =1, \ldots, M$):

$$\mathfrak{L}_{1i} \to \inf; \quad x_i(t_0) = x_0^i\,, \tag{10i}$$

$$\mathfrak{L}_{2j} \to \inf; \quad u_j(t) \in U_0^j\ \forall t \in [t_0, t_1]\,. \tag{11j}$$

Unter den obigen Voraussetzungen besitzen diese Aufgaben einen endlichen minimalen Zielfunktionswert, der auch angenommen wird. Aus den Lösungen der Probleme (10i) und (11j) wird dann der benötigte Subgradient zusammengesetzt; die Aufgaben selbst können mit beliebigen Methoden gelöst werden.

## 5. Einige numerische Erfahrungen

Der Algorithmus wurde u. a. an folgendem Beispiel erprobt[3]:

$$A=\left(\begin{array}{cc|c|c} 3 & -1 & & \\ -1 & 3 & & \\ \hline & & 3 & \\ \hline & & & 3 \end{array}\right), \quad B=\left(\begin{array}{c|c|c} 100 & & \\ \hline & 100 & \\ \hline & & 100 \end{array}\right),$$

$$C=\left(\begin{array}{cc|c|c} -0{,}2125 & 0{,}15 & 0{,}0625 & 0 \\ 0{,}2 & -0{,}2825 & 0 & 0{,}0825 \\ \hline 0{,}0925 & 0 & -0{,}0925 & 0 \\ \hline 0 & 0{,}1 & 0 & -0{,}1 \end{array}\right), \quad D=\left(\begin{array}{c|c|c} -1 & 0 & 0 \\ 0 & 1{,}33 & 0 \\ \hline 0 & -1{,}5 & 1{,}5 \\ \hline 1{,}33 & 0 & -1{,}33 \end{array}\right),$$

$t_0=0$, $t_1=25$, $x_0=(5,-5,0,0)^{\mathsf{T}}$. Eine Steuerbeschränkung liege nicht vor, d. h. $U_0=R^3$.

Die Dekomposition der Aufgabe wird durch die Linien angedeutet. Zur numerischen Lösung der Teilaufgaben wurde eine äquidistante Diskretisierung des Zeitintervalls $[t_0, t_1]$ mit $T$ Intervallen vorgenommen. Zur Einschätzung der Güte der mit dem Algorithmus berechneten Lösungen erfolgte ein Vergleich mit der für $T$ Diskretisierungsintervalle optimalen Lösung, die durch Diskretisierung der Ausgangsaufgabe erhalten wurde, was z. B. für $T=25$ auf die Lösung eines linearen Gleichungssystems mit 275 Unbekannten führte. Die Rechenzeit betrug für obiges Beispiel mit $T=25$ pro Iterationsschritt ca. 0,5 bis 0,7 s auf einer Anlage ES 1040. Abb. 1 zeigt die Trajektorien der dritten Zustandskomponenten für verschiedene Iterationen.

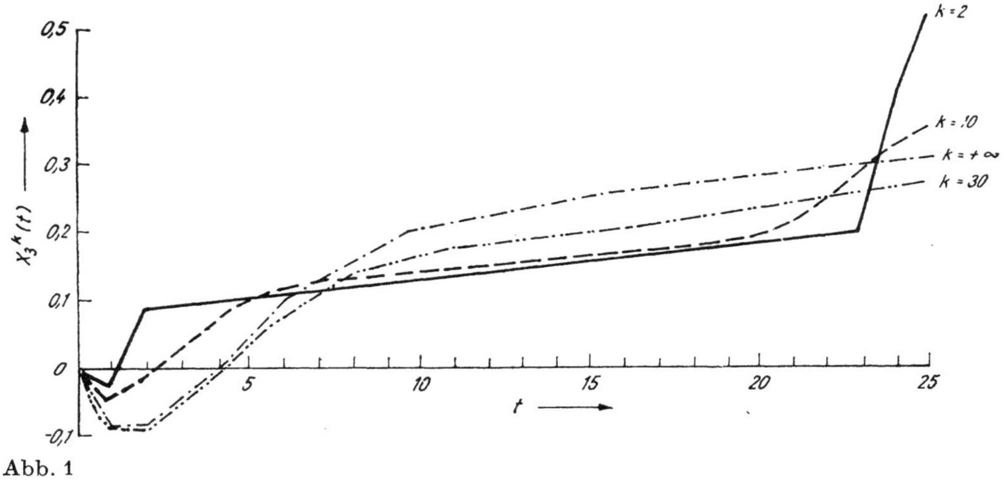


Abb. 1

[3] Für das sorgfältige Ausführen der Programmier- und Testarbeiten danke ich meinem Diplomanden, Herrn DIETER FRANKE.

Schrittweitenwahl:

$$\gamma_k = \frac{\lambda \cdot m_k}{\sqrt{\sqrt{k}+1}}, \quad m_k = \begin{cases} b > 1, & \text{wenn } \dfrac{\|a_{k-1}(\cdot)\|^2 - \|a_k(\cdot)\|^2}{\|a_k(\cdot)\|^2} < 0{,}2 \text{ und } k \equiv 0 \pmod 3, \\ c < 1, & \text{wenn } \|a_k(\cdot)\| > \|a_{k-1}(\cdot)\|, \\ 1, & \text{sonst.} \end{cases}$$

Hierbei ist $\lambda$ ein positiver variabler Parameter, der unter anderem von $T$ abhängig ist. Die Vergrößerung der Schrittweite durch $b$ soll einen erhöhten Zuwachs des Zielfunktionswertes sichern (es wurde z. B. $b = 10$ gewählt), während die Verminderung der Schrittweite durch $c$ verhindern soll, daß man sich zu sehr aus Optimumsnähe entfernt (z. B. $c = 0{,}5$ oder $c = 0{,}67$). Mit der angegebenen Schrittweitenwahl haben wir für die von uns untersuchten Beispiele bessere Erfahrungen gemacht als mit irgendeiner anderen Wahl der Schrittweite, wie sie in der Literatur vorgeschlagen wird (vgl. [1]).

Wahl der Ausgangsfunktion $p_0(\cdot)$: Die Güte der erzeugten Folge $\{p_k(\cdot)\}$ ist sehr davon abhängig, wie weit entfernt vom Optimum $p_*(\cdot)$ die Funktion $p_0(\cdot)$ liegt. Wir haben festgestellt, daß bei großer Entfernung vom Optimum auf den ersten 10 bis 20 Iterationen ein günstiges Verhalten bezüglich Zielfunktionswert und Subgradientennorm zu verzeichnen ist, während danach mitunter eine langsame Konvergenz erfolgt. Liegt $p_0(\cdot)$ in relativer Optimumsnähe, so ist das Konvergenzverhalten sehr gut.

Diese numerischen Erfahrungen lassen das Verfahren vor allem zur Ermittlung einer suboptimalen Lösung sehr geeignet erscheinen, die mit geringem Aufwand bereits nach wenigen Iterationen ermittelt werden kann.

Günstig erscheint uns auch folgende Modifikation des Verfahrens (vgl. [3]): Ausgehend von einer berechneten suboptimalen Steuerung $\bar{u}(\cdot)$ bestimme man über die (diskretisierte) Differentialgleichung $\dot{x} = Cx + D\bar{u}$ durch sukzessive Berechnung relativ leicht die zugehörige Phasentrajektorie $\bar{x}(\cdot)$. Berechnet man zu dem zulässigen Paar $(\bar{x}(\cdot), \bar{u}(\cdot)) = \bar{z}$ den Wert $\mathfrak{J}(\bar{x}(\cdot), \bar{u}(\cdot))$, so hat man zusammen mit dem aktuell besten Zielfunktionswert in der Dualaufgabe gleichzeitig eine Abschätzung für die Optimumsentfernung. Ist die erhaltene zulässige Lösung $\bar{z}$ noch nicht gut genug, so kann man aus $\bar{x}(\cdot)$ über die konjugierte Gleichung ein neues $\bar{p}(\cdot)$ ermitteln und als Anfangsfunktion für den Subgradientenalgorithmus verwenden, was sich bei den von uns untersuchten Beispielen außerordentlich vorteilhaft auswirkte.

## Literatur

[1] BEER, K.: Lösung großer linearer Optimierungsaufgaben. Dt. Verlag d. Wissenschaften, Berlin 1977.

[2] COHEN, G.: On an Algorithm of Decentralized Optimal Control. J. of Math. Analysis and Applic. **59** (1977), 242–259.

[3] FRANKE, D., LUDERER, B.: Realisierung eines Algorithmus zur von Aufgaben der optimalen Steuerung. WZ TH Karl-Marx-Stadt **7** (1979), 853–858.

[4] HADLEY, G.: Nichtlineare und dynamische Programmierung (Übers. a. d. Engl.). Verlag Die Wirtschaft, Berlin 1969.
[5] IOFFE, A. D., TICHOMIROV, V. M.: Theorie der Extremalaufgaben (Übers. a. d. Russ.). Dt. Verlag d. Wissenschaften, Berlin 1979.
[6] ISHIMATSU, T., MOHRI, A., TAKATA, M.: Optimization of Weakly Coupled Subsystems by a Two-level-method. Int. J. Control. **22** (1975) 6, 877–882.
[7] LUDERER, B.: Zu einigen Anwendungsmöglichkeiten von Dekompositionsprinzipien bei der Lösung von Optimalsteueraufgaben. WZ TH Karl-Marx-Stadt **6** (1978), 781–790.
[8] PEARSON, J. D.: Dynamic Decomposition Techniques. Ch. 4 in: Optimization Methods for Large-Scale Systems with Applications. Ed.: D. A. WISMER. McGraw-Hill, New York 1971.
[9] Поляк, Б. Т.: Один общий метод решения экстремальных задач. ДАН СССР **174** (1967) 1, 33–36.
[10] РОКАФЕЛЛАР, Р.: Выпуклый анализ (Перев. с англ.), « Мир », Москва 1973.
[11] ТАБАК, Д., и Б. КУО: Оптимальное управление и математическое программирование (Перев. с англ.). « Наука », Москва 1975.
[12] VAINIKKO, G.: Funktionalanalysis der Diskretisierungsmethoden. Teubner, Leipzig 1976.

## Summary

A method for solving convex optimization problems in HILBERT spaces, based on the principle of goal coordination, is described. This general approach is then applied to a linear-quadratic control problem, which is assumed to have a special structure. The arising infimal problems of smaller dimension and without dynamics must be solved several times. The method is illustrated by an example.

## Резюме

Описывается метод решения для выпуклых задач оптимизации в ГильБЕРТовых пространствах, основывающийся на двойственном подходе. Этот общий принцип потом применяется к линейно-квадратичной задаче оптимального управления с определенной структурой. При этом возникают вспомогательные задачи меньшей размерности и без ограничений. Статья кончается примером.

Received March 1979; revised January 1980

## Book Review

R. Trappl; G. J. Klir; L. Riccardi (Eds.): **Progress in Cybernetics and Systems Research.** Vol. III. John Wiley & Sons Lim., Chichester 1978, 674 S., 28,50.

Der vorliegende Band umfaßt unter den Themenkomplexen Allgemeine Systemtheorie, Biokybernetik und Theoretische Neurobiologie Beiträge anerkannter Vertreter der entsprechenden Wissensgebiete zusammen.

In den kurzen aber präzise gefaßten Arbeiten werden, durch die Auswahl und die Anordnung unterstützt, wesentliche Trends der entsprechenden Disziplinen vorgestellt und bewertet. Die Beiträge zur Systemtheorie verdeutlichen, daß mit den immer komplizierter werdenden Strukturen der technischen informationsverarbeitenden Systeme Methoden und Verfahren zur Beschreibung aktiver hierarchischer Mehrebenensysteme entwickelt und aufbereitet werden, derer die Biokybernetik dringend bedarf.

In dieser Hinsicht vermittelt das Buch einen guten Überblick über den derzeitigen Stand und aktuelle Probleme; werden für die Biokybernetik relevante Themen der allgemeinen Systemtheorie anwendungsorientiert diskutiert. Dabei werden sowohl grundlegende Begriffe wie Rolle der Zufälligkeit, Steuerbarkeit und Beobachtbarkeit, Adäquatheit von Beschreibung als auch spezielle Verfahren diskutiert.

Die Beiträge zur Biokybernetik und Theoretischen Neurobiologie sind thematisch relativ breit gefächert und reichen von der Neuroanatomie über die Beschreibung physiololgischer Systeme bis zu speziellen Problemen und Beispielen der theoretischen Modellierung in verschiedensten Gebieten der Biowissenschaften.

Das dokumentiert die enorme Entwicklung der Biokybernetik als Wissenschaftsdisziplin in den letzten Jahren, verdeutlicht aber auch die Schwierigkeiten, die ein derart heterogenes interdisziplinäres Gebiet für den Bearbeiter mit sich bringt.

Die gewählte Kombination grundsätzlicher Betrachtungen und spezieller Lösungsvorschläge verbunden mit umfangreichen bibliographischen Angaben erleichtert dem interessierten Leser das Einordnen der Arbeiten. Insofern ist diese Sammlung eine wertvolle Hilfe für das Einarbeiten in das Gebiet der Biokybernetik z. B. für Studenten im Spezialstudium der Bioingenieurwissenschaften sowie Interessenten aller Bereiche in der postgradualen Weiterbildung.

Für den Fachmann ist die Sammlung eine willkommene Ergänzung. Sie ragt durch Art und Auswahl der Arbeiten zum Nachdenken über Voraussetzungen und Bedingungen für die Anwendung der aus dem Bereich der technischen Kybernetik übernommenen Begriffe und Verfahren auf biologische Systeme und zur Berücksichtigung moderner Aspekte der Systemtheorie an. In diesem Sinne ist auch der mit 13 Beiträgen verhältnismäßig stark repräsentierte, der unscharfen Beschreibung gewidmete Abschnitt zu verstehen.

Andererseits fehlt das für die experimentelle Biokybernetik äußerst wichtige Gebiet der praktischen Systemidentifikation, also Identifikationsverfahren, aktive Versuchsplanung in bezug auf die Untersuchung biologischer Organismen.

E. Forth

Math. Operationsforsch. Statist., Ser. Optimization, Vol. 11 (1980) No. 4, 605–616

# Limited Risk Control of the ORNSTEIN-UHLENBECK Process

NORBERT CHRISTOPEIT[1] and KURT HELMES[2]

**Summary**: We consider the problem of steering a system desribed by a simple one dimensional linear stochastic differential equation in such a way as to maximize the upper allowable limit that is exceeded by the final state with a prescribed probability. The optimal controls are computed by applying the maximum principle to an aquivalent deterministic problem. This model can be used among other things to find optimal strategies for specific advertising activities of a firm.

## 1. Introduction

In [2] a control system is considered whose dynamics are governed by a linear stochastic differential equation

$$\mathrm{d}x = A(t)\,x(t)\,\mathrm{d}t + B(t)\,u(t)\,\mathrm{d}t + C(t)\,\mathrm{d}w, \quad 0 \leqq t \leqq T\,, \tag{1.1}$$

with initial condition

$$x(0) = x_0\,. \tag{1.2}$$

Here $w$ is a BROWNian motion defined on some probability space $(\Omega, \mathfrak{S}, \mathsf{P})$, and the matrices $A(t)$, $B(t)$ and $C(t)$ are nonstochastic. It is suggested that in certain situations (or for "risk averted" policy makers, see below) it may be advantageous to consider the control problem:

$$\begin{aligned} &\gamma = \max_{u \in \mathfrak{X}} \\ &\text{subject to (1.1), (1.2)} \\ &\text{and} \\ &\mathsf{P}(g(x(T)) \geqq \gamma) \geqq 1 - \alpha \end{aligned} \tag{P}$$

$(0 < \alpha < 1/2)$ instead of the standard problem

$$\begin{aligned} &\mathsf{E}(g(x(T))) = \max_{u \in \mathfrak{X}}! \\ &\text{subject to (1.1), (1.2).} \end{aligned} \tag{P1}$$

[1] Institut für Ökonomie und Operations Research, Universität Bonn, Adenauerallee 24–42, 53 Bonn, BRD.

[2] Institut für Angewandte Mathematik, Universität Bonn, Wegelerstr. 6, 53 Bonn, BRD.

As feasible controls $u \in \mathfrak{T}$ we admit all linear feedback controls of the form

$$\left.\begin{array}{l} u(t) = U(t)\, x(t) + v(t), \text{ where } U(\cdot), v(\cdot) \text{ are nonstochastic measurable} \\ \text{functions such that } (U(t), v(t)) \in \mathfrak{W} \text{ a.e., where } \mathfrak{W} \text{ is a subset} \\ \text{of a EUCLIDean space.} \end{array}\right\} \quad (1.3)$$

Here we shall consider the one dimensional controlled ORNSTEIN-UHLENBECK process

$$dx = -ax(t)\, dt + bu(t)\, dt + \varepsilon dw, \quad 0 \leqq t \leqq T , \quad (1.4)$$

with initial condition

$$x(0) = 0 . \quad (1.5)$$

The coefficients $a, b, \varepsilon$ are all assumed to be positive numbers. As *"performance index"* we take the final state itself:

$$g(x(T)) = x(T) . \quad (1.6)$$

The control functions $U(t)$, $v(t)$ (cf. (1.3)) are allowed to range in the intervals $[\underline{d}/b, \bar{d}/b]$ and $[0, v_{\max}]$ resp. and $a$, the coefficient of friction, lies in the open interval $(\underline{d}, \bar{d})$. These assumptions guarantee that a control function can be chosen so as to compensate the friction.

One of the reasons to investigate the problem (1.4)–(1.6) stems from the fact that this model can be used to describe the influence of specific marketing activities of a firm. When a new product, belonging to a class of products like cigarettes or washing detergents is introduced to the market, which for those kind of products usually has a fixed number of consumers, a large advertising campaign has to be prepared in advance. One objective is to create a positive view or remove a negative one.

The older marketing models are deterministic [cf. [1], ch. 7]. It is assumed that the so called *"stock of goodwill at time t"*, $y(t)$, satisfies the equation

$$\dot{y}(t) = -ay(t) + v(t) , \quad (1.7)$$

where it is assumed that the goodwill depreciates at a fixed rate $a$, but that it can be increased through advertising expenditures $v(t)$. The control function $v(t)$, satisfying some kind of restrictions, e.g. $0 \leqq v(t) \leqq v_{\max}$, is to be chosen such that a performance index, which incorporates the price and other important characteristics as well as $y(t)$ and $v(t)$, is maximized. While the model is very simple it provides reasonable policies in some special cases [cf. [1]].

Nevertheless there is no doubt that some parameters of the "system" and some of the parameters that are neglected are random and that their stochastic nature should be taken into account. This randomness leads to new questions, namely, what is a "good" objective function, or, should we use closed-loop controls $v(t, x)$ instead of open-loop controls.

Then, the advertising activity can depend on the present state $x(t)$ of the system (see 1.4 above) or even on its whole past.

In contrast to the way it is usually done in the literature the ultimate goal of a marketing strategy should not be to maximize the *expected* appreciation of a new product but to maximize the appreciation itself within a given probability. Since the success of a product-manager is intimately related to the success of the product, and maximizing the expectation does not exclude the possibility that one particular campaign might be a failure, we think that the performance index of our problem (P) provides a better description.

In contrast to the model (1.7), where $y(t)$ is assumed to be measured in "capital units", we interpret the variable $x(t)$ of equation (1.4) as measuring the attitude of one or all consumers to the new product. Its rate of change, its attraction, equals at least a sum of three different terms. The first one $(-ax(t))$ describes the effect of "losing the attraction" which could be caused by forgetting the product, thus reducing the magnitude of appreciation if it is negative. In [4] values of $a$ are estimated empirically. The third factor $(\varepsilon \mathrm{d}w(t))$ models the stochastic influence on our problem, originating either in the unknown "taste" of the consumers or in unforseen activities of competitors, etc. We think of these disturbances all together, to be "white noise". The second factor, which splits up into two terms, represents the advertising activity at time $t$. This splitting which appears in the feasible controls given by (1.3) occurs for two different reasons. The first reason is a mathematical one.

Since an admissible control is often a linear approximation to a non-linear feedback control which yields a more tractable problem. The second motive comes from the applications we have in mind. The attraction of a new product can be influenced by "form specific" advertising, which is assumed to be proportional to the state of the system $(bu(t)\,x(t))$, as well as by "class specific" advertising $(bv(t))$, where both control functions $u(t)$ and $v(t)$ are deterministic functions as the marketing strategy has to be fixed at the beginning of a planning period $T$.

We think that all these hypotheses are naturally satisfied if the new kind of the product reveals no "objective" improvements compared with competing products in the class and the price for all elements of the given product class is nearly the same. A by now classical example for such a situation is the advertising campaign which was prepared for the first menthol cigarette.

By a straightforward transformation of the state and control variables as well as of the parameters the model may be reduced to:

$$\mathrm{d}x = (u-a)\,x\mathrm{d}t + v\mathrm{d}t + \mathrm{d}w, \quad 0 \leqq t \leqq T\,,$$
$$x(0) = 0\,,$$

subject to the constraints:

$$0 \leqq u \leqq u_{\max}\,, \quad 0 \leqq v \leqq v_{\max}\,, \tag{P2}$$

$\gamma$, restricted to

$$\mathsf{P}(x(T) \geqq \gamma) \geqq 1-\alpha\,,$$

is to be maximized.

(By abuse of notation we have substituted "$u$" for "$U$" to indicate that we are dealing with the scalar case; this will not lead to confusion with the original $u$.)

Put $\mu(t)=\mathsf{E}(x(t))$, $\sigma^2(t)=\mathsf{var}\,(x(t))$, $\mathfrak{F}=\{(u, v) \mid [0, T]\to[0, u_{\max}]\times[0, v_{\max}]$ measurable$\}$ and $c=N_{1-\alpha}$, the $(1-\alpha)$ quantile of the standard normal distribution (note that $c<0$ for $\alpha<1/2$). Then—as shown in [2]—problem (P) for the special situation given by (P2) is equivalent to the (nonlinear) deterministic control problem.

$$\gamma=\mu(T)+c\sigma(T)=\max_{(u,v)\in\mathfrak{F}}!$$

subject to

$$\begin{aligned}&\dot{\mu}=(-a+u)\,\mu+v\,,\\&\dot{\overline{\sigma^2}}=2\,(-a+u)\,\sigma^2+1\,,\\&\mu(0)=0,\quad \sigma^2(0)=0\,.\end{aligned}\tag{P3}$$

To obtain a further simplification we observe that replacing the variables $\mu$, $v$, $\gamma$ and $c$ by $\mu/v_{\max}$, $v/v_{\max}$, $\gamma/v_{\max}$ and $c/v_{\max}$ we may assume $v_{\max}\equiv 1$ in (P3). In the sequel we shall compute optimal controls for (P3). It will turn out that they are in general different from those control laws which are optimal for the corresponding problem (P1).

## 2. The Structure of Extremal Controls

(P3) is a nonlinear control problem of Mayer type. Let us write down the necessary optimal conditions (cf. [3]):

Adjoint equations:

$$\dot{p}=-(-a+\bar{u})\,p\,,\tag{2.1}$$
$$\dot{q}=-2\,(-a+\bar{u})\,q\,.\tag{2.2}$$

Maximum principle:
for all $(u, v)\in[0, u_{\max}]\times[0, 1]$

$$\begin{aligned}&pv+q+(-a+u)\,p\mu+2\,(-a+u)\,q\sigma^2\\&\leqq p\bar{v}+q+(-a+\bar{u})\,p\mu+2\,(-a+\bar{u})\,q\sigma^2=\text{const}\end{aligned}\tag{2.3}$$

a.e. on $[0, T]$.

Transversality condition:

$$\begin{pmatrix}p(T)\\q(T)\end{pmatrix}=s\cdot\begin{pmatrix}1\\c/2\sigma(T)\end{pmatrix}\quad\text{for some}\quad s>0\,.\tag{2.4}$$

In accordance with common terminology a pair of controls $(\bar{u}, \bar{v})$ with corresponding responses (=solutions to the differential equations in (P3)) $(\mu, \sigma^2)$ will be called *extremal* if all of the conditions (2.1)–(2.4) are satisfied. From (2.4) it is clear that

$$p(T)>0,\quad q(T)<0\,,$$

hence, by (2.1) and (2.2),

$$p(t) > 0, \quad q(t) < 0 \quad \text{for all} \quad t\,, \tag{2.5}$$

and from (2.3)

$$\bar{v} \equiv 1\,.$$

Thus (2.3) reduces to

$$(\bar{u} - u)\,\varphi \geqq 0 \quad \text{a.e. on} \quad [0, T]\,,$$

where

$$\varphi = p\mu + 2q\sigma^2$$

is the switching function. It is plain to see that for a set $I \subset [0, T]$ of positive LEBESGUE measure

$$\varphi > 0 \quad \text{on } I \Rightarrow \bar{u} = u_{\max} \quad \text{on} \quad I \tag{2.6}$$

$$\varphi < 0 \quad \text{on } I \Rightarrow \bar{u} = 0 \quad \text{on} \quad I\,. \tag{2.7}$$

Less obvious is the singular case. It can be shown that (2.8)

$$\varphi = 0 \quad \text{on } I \Rightarrow \bar{u} = a \quad \text{on} \quad I\,.$$

Note that when (2.6)–(2.8) hold, all implications are also true in the converse direction. It will turn out that it suffices to verify (2.8) for intervals. Then $\varphi = 0$ on $I$ implies

$$\dot{\varphi} = p + 2q = 0$$
$$\ddot{\varphi} = \dot{p} + 2\dot{q} = -(-a + \bar{u})\,p - 4\,(-a + u)\,q = 0$$

on $I$. By virtue of (2.5) these two equations can hold simultaneously only if $\bar{u} = a$.

Next we show that once $\varphi$ is positive it remains so. To this end, suppose that $\varphi(\bar{t}) = 0$ and $\varphi > 0$ on $(\bar{t}, \bar{t} + \varepsilon)$ for some $\varepsilon > 0$. Then $\varphi$ must be increasing in some right neighborhood of $\bar{t}$, i.e.

$$\dot{\varphi} = p + 2q \geqq 0 \quad \text{on} \quad (\bar{t}, \bar{t} + \delta)$$

for some $0 < \delta < \varepsilon$. Hence, for arbitrary points $t_1 \in (\bar{t}, \bar{t} + \delta)$, $t_2 \in (\bar{t}, \bar{t} + \varepsilon)$, $t_2 > t_1$, and denoting by $\varkappa$ the difference $(u_{\max} - a)$,

$$\begin{aligned}\dot{\varphi}(t_2) = p(t_2) + 2q(t_2) &= [p(t_1) + 2q(t_1)\, e^{-\varkappa(t_2 - t_1)}] \times e^{-\varkappa(t_2 - t_1)} \\ &\geqq [p(t_1) + 2q(t_1)]\, e^{-\varkappa(t_2 - t_1)} \\ &\geqq 0\,.\end{aligned}$$

The same sort of argument applies to the case when $\varphi$ becomes negative (with $k$ replaced by $-a$). Moreover, this argument shows that the region where $\varphi$ vanishes must be an interval with left endpoint 0.

So we find that only controls of a very special structure can be extremal:

**Theorem 1:** *Any extremal control $\bar{u}$ must be of the form*

$$\bar{u} = a \quad \textit{on} \quad [0, \tau], \quad \bar{u} = 0 \quad \textit{or} \quad \bar{u} = u_{\max} \quad \textit{on} \quad (\tau, T]\,.$$

## 3. Computation of the Optimal Control

We begin with some lemmas providing additional insight into the properties of extremal controls. Consider any extremal control switching from $a$ to $u_{\max}$ at some time $\tau \in [0, T]$. The corresponding solution of the differential equations in (P3) for $t \geqq \tau$ are given by

$$\mu(t) = e^{\varkappa(t-\tau)} \left[\tau + \int_\tau^t e^{-\varkappa(s-\tau)}\, ds\right] = e^{\varkappa(t-\tau)} \left(\tau + \frac{1}{\varkappa}\right) - \frac{1}{\varkappa}, \tag{3.1}$$

$$\sigma^2(t) = e^{2\varkappa(t-\tau)} \left[\tau + \int_\tau^t e^{-2\varkappa(s-\tau)}\, ds\right] = e^{2\varkappa(t-\tau)} \left(\tau + \frac{1}{2\varkappa}\right) - \frac{1}{2\varkappa} \tag{3.2}$$

with $\varkappa = u_{\max} - a$, and the corresponding value $\bar{y}$ of the problem is given by

$$\bar{y} = \mu(T) + c\sigma(T) . \tag{3.3}$$

On the other hand, it follows from the transversality condition that

$$\sigma(T) = \frac{c}{2} \frac{p(T)}{q(T)} = \frac{c}{2} \pi e^{\varkappa \Delta}, \tag{3.4}$$

where we have put

$$\pi = \frac{p(0)}{q(0)}, \quad \Delta = T - \tau .$$

Together with (3.2) this implies

$$e^{2\varkappa\Delta} (1 - \beta\pi^2 + 2\varkappa\tau) = 1 , \tag{3.5}$$

with $\beta = \varkappa c^2/2$.

**Lemma 1:** *If an extremal control switches from $a$ to $u_{\max}$ at some time $\tau < T$, then the corresponding value $\bar{y}_{a|u_{\max}}$ of (P3) is positive. To the control $\bar{u} \equiv a$ (i.e. $\tau = T$) corresponds the value $\bar{y}_a = 0$.*

Proof: For any extremal control, let $\bar{y}$ denote the corresponding level. Then

$$\mu(T) = \bar{y} - c\sigma(T) = \bar{y} - \frac{c^2}{2} \frac{p(T)}{q(T)}. \tag{3.6}$$

Substituting (3.4) and (3.6) into

$$\varphi(T) = \mu(T)\, p(T) + 2q(T)\, \sigma^2(T)$$

yields

$$\varphi(T) = \bar{y} p(T) .$$

Since $p(T) > 0$ and $\varphi(T) > 0$ $[\varphi(T) = 0]$ for $\tau < T$ $[\tau = T]$, it follows that

$$\bar{y}_{a|u_{\max}} > 0 \quad \text{and} \quad \bar{y}_a = 0 . \quad \blacksquare$$

**Lemma 2:** *The control $\bar{u} \equiv u_{\max}$ is extremal if and only if*

$$4\beta \leqq 1 - e^{-2\varkappa T} . \tag{3.7}$$

Proof: We shall begin by proving that for $u \equiv u_{\max}$ extremality is equivalent to:

(A) $e^{2\varkappa T}(1-\beta\pi^2) = 1$ has a solution $\pi \leqq -2$.

To show sufficiency, define the adjoint variables $p, q$ as the solutions of (2.1), (2.2) with $p(T) = s > 0$ and $q(0) = p(0)/\pi$. Since (A) together with the solution formula (3.2) for $\sigma^2$ (with $\tau = 0$) implies

$$\sigma(T) = \frac{c}{2} \pi e^{\varkappa T} ,$$

the transversality condition follows from

$$q(T) = \frac{p(0)}{\pi} e^{-2\varkappa T} = \frac{p(T)}{\pi} e^{-\varkappa T} .$$

Next consider the switching function

$$\varphi(t) = q(t)\, \sigma^2(t) \left[ 2 + \frac{p\mu}{q\sigma^2}(t) \right] . \tag{3.8}$$

In order to show (2.6)–(2.8), we have to verify that $\varphi(t) > 0$ for $t > 0$, or, equivalently,

$$\psi(t) = \frac{p\mu}{q\sigma^2}(t) < -2 . \tag{3.9}$$

But

$$\psi(t) = \pi e^{\varkappa t} \frac{(e^{\varkappa t}-1)/\varkappa}{(e^{2\varkappa t}-1)/2\varkappa} = \frac{2\pi}{1+e^{-\varkappa t}} ;$$

hence, taking $t \downarrow 0$, (3.9) is equivalent to

$$\pi \leqq -2 .$$

Necessity of (A) follows from this last equivalence and the derivation of (3.5) (with $\tau = 0$) from (3.2) and (3.4).

It remains to show the equivalence of (A) and (3.7). By a simple calculation, (A) implies (3.7). Conversely if (3.7) holds,

$$\pi = -\sqrt{\frac{1}{\beta}(1 - e^{-2\varkappa T})}$$

will satisfy (A). ∎

**Lemma 3:** *There exists an extremal control $u_{a|u_{\max}}$ switching from $a$ to $u_{\max}$ at some time $\tau > 0$ if and only if*

$$c^2 \leqq T \quad \textit{and} \quad 4\beta > 1 - e^{-2\varkappa T} .$$

Proof: Let $u_{a|u_{\max}}$ be extremal. Looking at the switching function (3.8) we see that $\varphi(t)=0$ on the interval $[0, \tau]$ implies $p\mu/q\sigma^2 \equiv -2$. Hence

$$\lim_{t\downarrow 0} \frac{p\mu}{q\sigma^2}(t)=\lim_{t\downarrow 0} \frac{\dot{\overline{p\mu}}}{\dot{\overline{q\sigma^2}}}(t)=\lim_{t\downarrow 0} \frac{p(t)}{q(t)}=\pi=-2\ .$$

Then (3.5) takes the form

$$1-4\beta+2\varkappa\tau=e^{-2\varkappa\Delta} \tag{3.10}$$

or

$$\log(1-4\beta+\tau')=-\Delta',$$

where we have introduced the transformed time scale

$$t'=2\varkappa t\ .$$

Put

$$g(t')=\log(1-4\beta+t')\ .$$

Then $g(t')$ is defined for all $t'>4\beta-1$. We are looking for an intersection point $\tau'$ of $g(t')$ and $-\Delta'$ in the interval $[0, T']$. To this end, observe first that the root of $g(t')$ is

$$t_0'=4\beta$$

and that at this and only at this point $g(t')$ has slope 1. Now distinguish two cases: $4\beta\geqq 1$ and $4\beta<1$. In the first case (see fig. 1), a straightforward application of the meanvalue theorem together with the strict monotonicity of $g(t')$ shows that a unique intersection point $\tau'\in[0, T']$ exists if and only if the root lies in $[0, T']$, equivalently, if and only if

$$c^2\leqq T\ .$$

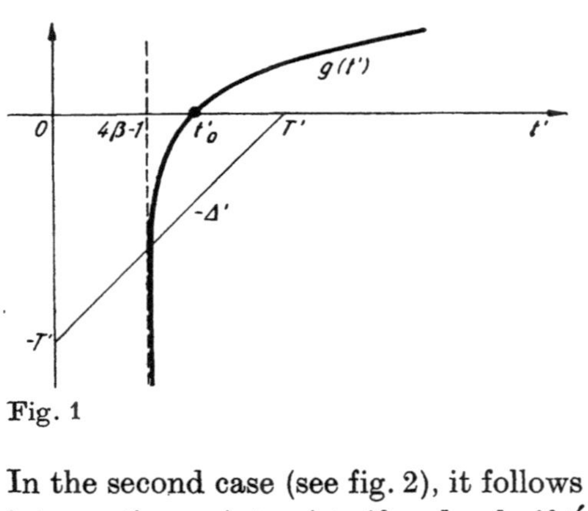


Fig. 1

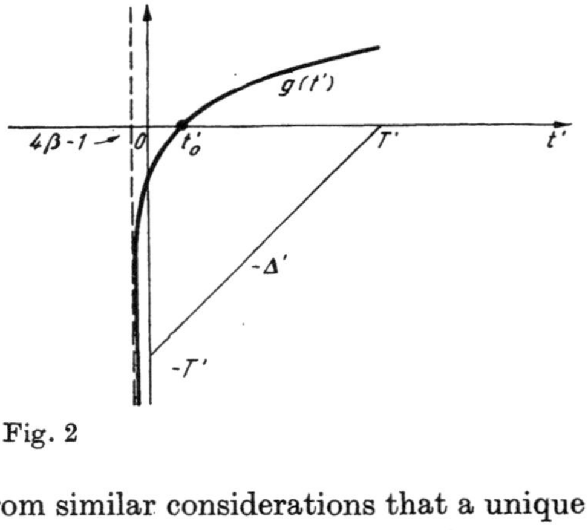


Fig. 2

In the second case (see fig. 2), it follows from similar considerations that a unique intersection point exists if and only if $t_0'\leqq T'$ and $g(0)<-T'$, equivalently,

$$c^2\leqq T \quad \text{and} \quad 4\beta>1-e^{-2\varkappa T}\ . \tag{3.11}$$

Note that the last condition is automatically satisfied in the case $4\beta\geqq 1$.

Hence (3.11) is necessary and sufficient for (3.10) to hold with $\tau\in[0, T]$, and hence necessary for $u_{a|u_{max}}$ to be extremal.

As for sufficiency, consider $u_{a|u_{\max}}$ switching from $a$ to $u_{\max}$ at time $\tau$, where $\tau$ is the (unique) solution of (3.10). Proceeding as in the proof of lemma 2, define adjoint variables $p, q$ as the solutions of (2.1), (2.2) with $p(T)=s>0$ and $q(0)=-p(0)/2$. Since (3.10) together with (3.2) implies

$$\sigma(T)=-ce^{\varkappa\Delta},$$

the transversality condition follows from

$$q(T)=-\frac{p(0)}{2}\,e^{-2\varkappa\Delta}=-\frac{p(T)}{2}\,e^{-\varkappa\Delta}.$$

Moreover, for $0\leqq t\leqq\tau$,

$$\varphi(t)=q(0)\,t\left[2+\frac{p(0)}{q(0)}\right]=0,$$

and, for $t>\tau$,

$$\frac{p\mu}{q\sigma^2}(t)=-2e^{\varkappa\Delta}\,\frac{e^{\varkappa\Delta}\left(\tau+\frac{1}{\varkappa}\right)-\frac{1}{\varkappa}}{e^{2\varkappa\Delta}\left(\tau+\frac{1}{2\varkappa}\right)-\frac{1}{2\varkappa}}$$

$$=-2\,\frac{2\varkappa\tau+1-e^{-\varkappa\Delta}+(1-e^{-\varkappa\Delta})}{2\varkappa\tau+1-e^{-\varkappa\Delta}+e^{-\varkappa\Delta}(1-e^{-\varkappa\Delta})}$$

$$<-2,$$

hence

$$\varphi(t)>0$$

for all such $t$. This shows (2.6)–(2.8). ■

Finally, consider controls $u_{a|0}$ switching from $a$ to 0 at some time $\tau\in[0, T]$. The solution formulae are given by (3.1), (3.2) with $\varkappa$ replaced by $-a$, and (3.5) becomes

$$e^{-2a\Delta}\,(1+\beta'\pi^2-2a\tau)=1, \tag{3.12}$$

where now $\beta'=ac^2/2$. As with $u_{a|u_{\max}}$, $\pi=-2$ if $u_{a|0}$ $(\tau>0)$ is extremal, thus leading to the equation

$$1+4\beta'-2a\tau=e^{2a\Delta} \tag{3.13}$$

for the determination of the switching time. By doing a completely analogous analysis as in the proof of lemma 3 the following result can be shown.

**Lemma 4.** (3.13) *determines a (unique) switching time* $0<\tau<T$ *for an extremal control* $u_{a|0}$ *switching from* $a$ *to* 0 *if and only if*

$$c^2>T \quad \textit{and} \quad 4\beta'<e^{2aT}-1.$$

The case $4\beta'=e^{2aT}-1$ is easily seen to correspond to $\tau=0$. The proof of the following lemma is along the same lines as that of lemma 1.

**Lemma 5.** *If an extremal control switches from a to 0 at some time* $\tau < T$, *then the corresponding value* $\bar{y}_{a|0}$ *is negative.*

Finally, $u \equiv 0$ extremal implies $\varphi(t) < 0$ for all $t > 0$, hence, by (3.9), $\psi(t) > -2$ and $0 \geqq \pi \lim_{t \searrow 0} \psi(t) \geqq -2$ (cf. the proof of lemma 3). Hence, from (3.12) (with $\tau = 0$):

**Lemma 6.** *If* $u \equiv 0$ *is extremal then* $4\beta' \geqq e^{2aT} - 1$.

Putting together all the lemmas, we obtain:

**Theorem 2.** *The optimal control for* (P3) *in its dependence on the parameters is given by the following diagram:*

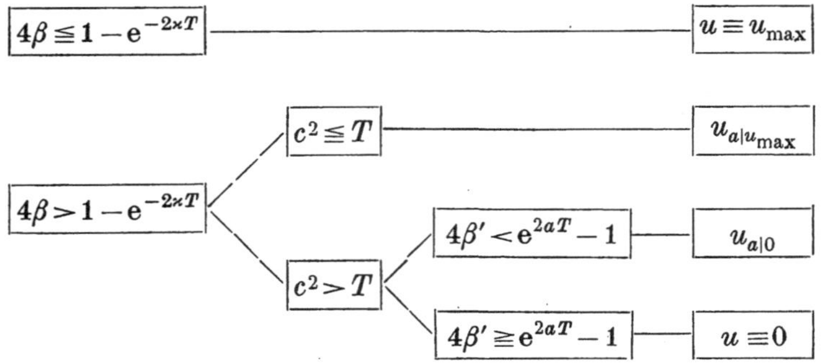


*The switching times for the controls* $u_{a|u_{max}}$ *and* $u_{a|0}$ *are computed from* (3.10) *and* (3.13), *respectively. The value* $\bar{y}$ *is computed from* (3.3) *in any case.*

## 4. Discussion of the Results

In order to discuss the results obtained in Th. 2 we have to go back to the original model (1.4) since the transformations obscure the influence of the initial parameters $\varepsilon, \varkappa, \tilde{v}_{max}$ and $T$ in the choice of the optimal control. Tracing back the transformation in the introduction we find that the parameter $c^2$ is actually given by

$$c^2 = c_\alpha^2 \cdot \varepsilon^2/\tilde{v}_{max} = c_\alpha^2/r \,. \tag{4.1}$$

($c_\alpha$ is the $(1-\alpha)$ quantile of the $N(0,1)$-distribution.) In (4.1) we have introduced the external impact rate $r$ which is the square of the ratio of the greatest input and the stochastic influence. Decisive for the choice of the optimal control is, however, not the rate $r$ itself but the total external impact $R$ during the planning period $T$, which is $R = rT$, combined with the feedback impact $K = \varkappa T$ and the friction impact $A = aT$. Using the intrinsic parameters $R, K, A$ and $c_\alpha^2$ of the optimal control problem (1.4), Th. 2 can be written as:

$$u_{opt} = \begin{cases} u \equiv u_{max} & \text{if} \quad c_\alpha^2 \cdot 2K/(1 - e^{-2K}) \leqq R \\ u_{a|u_{max}} & \text{if} \quad c_\alpha^2 \leqq R < c_\alpha^2 \cdot 2K/(1 - e^{-2K}) \\ u_{a|0} & \text{if} \quad c_\alpha^2 \cdot 2A/(e^{2A} - 1) < R < c_\alpha^2 \\ u \equiv 0 & \text{if} \quad R \leqq c_\alpha^2 \cdot 2A/(e^{2A} - 1)\,. \end{cases} \tag{4.2}$$

Hence the control $u \equiv u_{max}$ is optimal only if the external impact $R$ is "large enough", which is true if the "noise" (uncertainly) is small ($\varepsilon \ll 1$) and/or the maximal value of the advertising activity $v$ is large ($\tilde{v}_{max} \ll 1$). This is plausible since then the chance of getting a negative $x$ value and hence the risk of intensifying an undesired trend by the impact of $u$ is rather small. On the other hand if the external impact $R$ is low compared to the prescribed level $\alpha$ the control is cut off after some time $t$ to let the friction stabilize the system. In this case the optimal value $\gamma$ is always negative (cf. lemma 5). Therefore one of the objectives of a product manager should always be to ensure that the external impact $R$ of the campaign is sufficiently large. In the case between the two extreme situations described above the optimal control $u_{a|u_{max}}$ yields a positive value (cf. lemma 1), and the form of the optimal control makes sense since the switch from $a$ to $u_{max}$ takes place when we are sure to be "on the sunny side of the street" ($x$ is positive with sufficiently large probability).

It is interesting to compare this result with the solution of the standard problem (P1). If the linear feedback controls (1.3) are admitted as feasible control laws, it is easy to see that (P1) is equivalent to the deterministic control problem

$$\mu(T) = \max_{(u,v)\in\mathfrak{F}} !$$

subject to

$$\dot{\mu} = (-a+u)\,\mu + v\,.$$

A straightforward application of the maximum principle shows that $\bar{u} \equiv u_{max}$, $\bar{v} \equiv 1$ is the optimal solution. Hence the two solutions coincide only for certain regions in the parameter space.

## References


[1] BENSOUSSAN, A. et alii: Management Applications of Modern Control Theory. North Holland, Amsterdam 1974.

[2] CHRISTOPEIT, N.: A Limited Risk Model in Stochastic Control. Operations Research Verfahren XXVIII, 1978, 145–152.

[3] LEE, E. B. and L. MARKUS: Foundation of Optimal Control Theory. Wiley, New York 1967.

[4] TELSER, L. G.: Advertising and Cigarettes. J. Political Economy **70**, (1962) 5, 471–499.


## Zusammenfassung


Es wird das Problem betrachtet, ein System, das durch eine einfache lineare stochastische Differentialgleichung beschrieben wird, so zu steuern, daß der größte zulässige Wert, der vom Endzustand mit vorgegebener Mindestwahrscheinlichkeit überschritten wird, maximal wird. Die Berechnung der optimalen Kontrollen geschieht durch Anwendung des Maximumprinzips auf ein äquivalentes deterministisches Problem. Dieses Modell kann unter anderem zur Bestimmung optimaler Strategien bei speziellen Werbekampagnen von Firmen dienen.

**Резюме**

Рассматривается задачу управлять систему описываемую при помощи одномерного линейного стохастического дифференциального управления так что найбольшее допустимое значение превышаемое с заданной вероятностью конечным состоянием становится максимальным. Вычисление оптимальных управлений производится применением принципа максимума на эквивалентную неслучайную задачу. Эта модель можно применить например для вычисления оптимальных стратегий в специальных вербовых кампаниях от фирм.

Received May 1978; revised January 1980

Math. Operationsforsch. Statist., Ser. Optimization, Vol. 11 (1980) No. 4, 617–622

# A 2-unit System with "Imperfect" Repair

R. SUBRAMANIAN [1] and R. NATARAJAN [1]

**Summary:** In the study of repairable systems, it is usually assumed that a unit is 'new' after repair. While this assumption simplifies the analysis of such systems, there are many instances, in which this assumption is not true. In this paper, we study a 2-unit cold standby redundant system, in which the number of repairs that can be done for each unit is $k$ and after each repair a unit has a different failure time distribution.

## 1. Introduction

Repair maintenance and standby redundancy are well known methods by which the operating characteristics of a system can be improved [1, 2]. Redundant repairable systems have been studied extensively by several authors in the past. A bibliography of work on 2-unit systems has been given by OSAKI and NAKAGAWA [3]. In all the models considered so far, it is assumed that the units in the system are 'new' after repair and they can be repaired as many times as needed. While these assumptions simplify the analysis, there are many instances where they are not true. Thus it would appear that in a more realistic and reasonable model we have to assume different failure time distributions for the units after each repair. It is intuitively clear that a unit will not perform the desired operation satisfactorily after it has gone through 'sufficient' number of repairs. Hence it is necessary to put a restriction on the number of repairs a unit can undergo. The present contribution incorporates these aspects in the study of 2-unit redundant repairable systems.

In this paper we consider a two-unit cold standby redundant system with the following properties:

(i) The two units are dissimilar. We label them as unit 1 and unit 2. The two units have equal status: i.e., any unit performs the function of the system, even though their statistical properties are different.

(ii) All switchover times are negligible; switching is perfect.

(iii) Initially both the units are good; unit 1 is put online and the other is kept as cold standby.

(iv) There is only one repair facility. The repair is undertaken in FIFO order.

[1] Department of Mathematics, Indian Institute of Technology, Madras 600 036, India.

(v) The repair is 'imperfect' in the sense that the repaired unit will not behave like a 'new' one. The failure time distribution of either unit is different after each repair such that its expected life decreases after the repair.

(vi) Each unit can be repaired only $k$ times. When a unit fails for the $(k+1)$th time it cannot be repaired further.

## 2. Notations

$f_{i,n}(\cdot)$, $F_{i,n}(\cdot)$, $\bar{F}_{i,n}(\cdot)$ = pdf, cdf, sf of failure time of unit $i$ when it operates online for the $n$-th time, $i=1, 2$.

$g_i(\cdot)$, $G_i(\cdot)$, $\bar{G}_i(\cdot)$ = pdf, cdf, sf of the repair time of unit $i$, $i=1, 2$.

$f(t) * g(t)$ = convolution of the functions $f(t)$ and $g(t)$

$\psi^*(s)$ = LAPLACE transform of an arbitrary function $\psi(t)$

$A(t)$ = pr {the system is available at $t$}

$R(t)$ = pr {the system is operable in $(0, t)$}

## 3. Availability Analysis

We define the following events required for the analysis.

$E_{i,0}$ event that unit $i$ is new and just begins to operate online, $i=1, 2$.

$E_{i,j}$ event that unit $i$ just begins to operate online after its $j$-th repair, $i=1, 2$; $j=1, 2, \ldots k$.

Also let

$$p_a(n, t)\,\Delta = \mathrm{pr}\,\{E_{2,n} \text{ occurs between } t \text{ and } (t+\Delta) \mid E_{1,n} \text{ at } t=0\}, \quad n=1, 2, \ldots k$$

$$q_n(n, t)\,\Delta = \mathrm{pr}\,\{E_{1,n+1} \text{ occurs between } t \text{ and } (t+\Delta) \mid E_{2,n} \text{ at } t=0\}, \quad n=1, 2, \ldots k-1.$$

Since the repair is undertaken in the FIFO order unit 1 and unit 2 will operate online alternately. Initially unit 1 operates online and unit 2 is kept as standby. Hence unit 2 will begin to operate online after its $n$-th repair, upon the $(n+1)$th failure of unit 1. Thus

$$p_a(n, t) = f_{1,n+1}(t)\, G_2(t) + g_2(t)\, F_{1,n+1}(t)$$
$$q_a(n, t) = f_{2,n+1}(t)\, G_1(t) + g_1(t)\, F_{2,n+1}(t)\,.$$

If

$$\Phi_a(n, t)\,\Delta = \mathrm{pr}\,\{E_{1,n} \text{ between } t \text{ and } (t+\Delta) \mid E_{1,n-1} \text{ at } t=0\}, \quad n=1, 2, \ldots k$$

then it follows that

$$\Phi_a(n, t) = p_a\,(n-1, t) * q_a\,(n-1, t) \quad \text{for} \quad n=2, 3, \ldots k$$
$$\Phi_a(1, t) = f_{1,1}(t) * \{f_{2,1}(t)\, G_1(t) + F_{2,1}(t)\, g_1(t)\}\,.$$

The functions $\pi_a(n, t)$ defined by

$$\pi_a(n, t)\, \Delta = \text{pr}\, \{E_{1,n} \text{ between } t \text{ and } (t+\Delta) \mid E_{1,0} \text{ at } t=0\}, \quad n=1, 2, \ldots k$$

are determined from the following recurrence relations

$$\pi_a(n, t) = \pi_a\,(n-1, t) * \Phi_a(n, t) \quad \text{for} \quad n=2, \ldots k$$

$$\pi_a(1, t) = \Phi_a(1, t).$$

We are now in a position to write an expression for the availability of the system. Observing that the point where unit 1 is put online after repair is a point of regeneration and by considering the following mutually exclusive and exhaustive cases

(i) unit 1, which is new at $t=0$ may fail after $t$
(ii) unit 1 fails before $t$, unit 2 which is instantaneously put online fails after $t$
(iii) unit $i$ ($i=1, 2$) while operating online for the $n$-th time ($n \geqq 2$) fails after $t$
we have

$$A(t) = \bar{F}_{1,1}(t) + f_{1,1}(t) * \bar{F}_{2,1}(t) + \sum_{n=1}^{k} \pi_a(n, t) * \{\bar{F}_{1,n+1}(t) + p_a(n, t) * \bar{F}_{2,n+1}(t)\}.$$

## 4. Reliability Analysis

We define

$$p_r(n, t)\, \Delta = \text{pr}\, \{E_{2,n} \text{ occurs between } t \text{ and } (t+\Delta) \text{ and the system is operable in } (0, t) \mid E_{1,n} \text{ at } t=0\}, \quad n=1, 2, \ldots k$$

$$q_r(n, t)\, \Delta = \text{pr}\, \{E_{1,n+1} \text{ occurs between } t \text{ and } (t+\Delta) \text{ and the system is operable in } (0, t) \mid E_{2,n} \text{ at } t=0\}, \quad n=1, 2, \ldots k-1$$

$$\Phi_r(n, t)\, \Delta = \text{pr}\, \{E_{1,n} \text{ occurs between } t \text{ and } (t+\Delta) \text{ and the system is operable in } (0, t) \mid E_{1,n-1} \text{ at } t=0\}, \quad n=1, 2, \ldots k$$

$$\pi_r(n, t)\, \Delta = \text{pr}\, \{E_{1,n} \text{ occurs between } t \text{ and } (t+\Delta) \text{ and the system is operable in } (0, t) \mid E_{1,0} \text{ at } t=0\}, \quad n=1, 2, \ldots k.$$

Proceeding on lines similar to those in the previous section, we get

$$p_r(n, t) = f_{1,n+1}(t)\, G_2(t)$$

$$q_r(n, t) = f_{2,n+1}(t)\, G_1(t)$$

$$\Phi_r(n, t) = p_r\,(n-1, t) * q_r\,(n-1, t)$$

$$\Phi_r(1, t) = f_{1,1}(t) * f_{2,1}(t)\, G_1(t)$$

$$\pi_r(n, t) = \pi_r\,(n-1, t) * \dot{\Phi}_r(n, t)$$
$$\pi_r(1, t) = \Phi_r(1, t)$$
$$R(t) = \bar{F}_{1,1}(t) + f_{1,1}(t) * \bar{F}_{2,1}(t) + \sum_{n=1}^{k} \pi_r(n, t) * \{\bar{F}_{1,n+1}(t) + p_r(n, t) * \bar{F}_{2,n+1}(t)\}.$$

## 5. Illustrative Example

For the purpose of illustration, we consider a model in which both the units are identical, and their repair times are random variables with the same distribution function. For the sake of simplicity, we further assume that all the concerned distributions are exponential:

$$f_{1,n}(t) = f_{2,n}(t) = \lambda_n e^{-\lambda_n t}, \quad n = 1, 2, \ldots k+1$$
$$g_1(t) = g_2(t) = \mu e^{-\mu t}.$$

Then we have

$$A^*(s) = \frac{1}{s+\lambda_1}\frac{1}{(s+\lambda_1)^2} + \sum_{n=1}^{k} \frac{1}{s+\lambda_1} \frac{\lambda_1\mu\,(2s+\lambda_1+\mu)}{(s+\lambda_1)\,(s+\mu)\,(s+\lambda_1+\mu)} \frac{\mu^{2(n-1)}}{(s+\mu)^{2(n-1)}} \times \prod_{j=2}^{n} \left\{\frac{\lambda_j\,(2s+\lambda_j+\mu)}{(s+\lambda_j)\,(s+\lambda_j+\mu)}\right\}^2 \cdot \frac{1}{s+\lambda_{n+1}} \times \left\{1 + \frac{\lambda_{n+1}\mu\,(2s+\lambda_{n+1}+\mu)}{(s+\lambda_{n+1})\,(s+\mu)\,(s+\lambda_{n+1}+\mu)}\right\}$$

$$R^*(s) = \frac{1}{s+\lambda_1} + \frac{\lambda_1}{s+\lambda_1}\frac{1}{s+\lambda_1} + \sum_{n=1}^{k} \frac{\lambda_1}{s+\lambda_1} \frac{\lambda_1\mu}{(s+\lambda_1)\,(s+\lambda_1+\mu)}\mu^{2(n-1)} \times \prod_{j=2}^{n} \left\{\frac{\lambda_j}{(s+\lambda_j)\,(s+\lambda_j+\mu)}\right\}^2 \left\{\frac{1}{s+\lambda_{n+1}} + \frac{\lambda_{n+1}\mu}{(s+\lambda_{n+1})\,(s+\lambda_{n+1}+\mu)}\frac{1}{s+\lambda_{n+1}}\right\}$$

The meantime to first system failure is given by

$$R^*(0) = \frac{2}{\lambda_1} + \frac{1}{1+\varrho_1}\frac{1}{\lambda_2}\left(\frac{2+\varrho_2}{1+\varrho_1}\right) + \sum_{n=1}^{k} \frac{1}{1+\varrho_1} \prod_{j=2}^{n} \left(\frac{1}{1+\varrho_j}\right)^2 \frac{1}{\lambda_{n+1}} \left(\frac{\varrho_{n+1}+2}{\varrho_{n+1}+1}\right)$$

where

$$\varrho_j = \frac{\lambda_j}{\mu}.$$

In particular if we take $\lambda_n = n\lambda$ where $\lambda = 0.01$, and $\mu = 0.01$, we get the following results:

| $k$ | Meantime to first system failure |
|---|---|
| 10 | 602.83 |
| 20 | 724.78 |
| 30 | 796.15 |
| 40 | 844.56 |
| 50 | 879.48 |
| 100 | 958.87 |
| 200 | 979.43 |
| 300 | 980.00 |
| 400 | 980.00 |
| 500 | 980.00 |

The above table clearly indicates that it is not worthwhile to repair the units more than 200 times as the increase in the mean time to first system failure is negligible by further repair.

We also observe, that $A^*(s)$ and $R^*(s)$ are rational functions of $s$, and hence they can be easily inverted if $k$ is small and thus the availability and the reliability can be explicitly computed.

## References

[1] Barlow, R. E.; F. Proschan: Mathematical Theory of Reliability, John-Wiley, 1965.

[2] Gnedenko, B. V.; Yu. K. Belyayev and A. D. Solovyev: Mathematical Methods of Reliability Theory. Academic Press, New York, 1969.

[3] Osaki, S.; T. Nakagawa: Bibliography for reliability and availability of stochastic systems. IEEE Trans. Rel. R-**25** (1976) 4, 284–287.

## Zusammenfassung

Bei der Untersuchung von Erneuerungssystemen wird gewöhnlich vorausgesetzt, daß die Einheit nach der Reparatur „neu" ist.

Während diese Voraussetzung die Analyse solcher Systeme vereinfacht, gibt es jedoch viele Beispiele, in denen diese Voraussetzung nicht erfüllt ist.

In dieser Arbeit untersuchen wir ein doubliertes System mit kalter Reserve, in dem die Anzahl der Erneuerungen, die für jede Einheit durchgeführt werden können, gleich $k$ ist und eine Einheit nach jeder Erneuerung eine andere Verteilung der Ausfallzeit besitzt.

**Резюме**

При исследовании систем обновления обычно предпологается, что прибор после ремонта является « новым ». Хотя это предпосылка облегчает анализ таких систем, существуют много примеров, в которых она не выполняется. В настоящей работе рассматриваем дублированную систему с холодным резервом, в которой число обновлений возможны для каждого прибора равно $K$ и после каждого обновления прибор обладает другим распределением времени отказа.

Received October 1978, revised July 1979

Math. Operationsforsch. Statist., Ser. Optimization, Vol. 11 (1980) No. 4, 623–628

# A Single Unit Cumulative Damage Model with Correlated Damage and Repair Time

A. SUBRAMANIAN [1] and N. EAGAMBARAM [1]

**Summary:** In this paper we study the correlation between the life time and the repair time of a device which is subjected to a damage process. The correlation is obtained in section 2 when there is a repair facility which repairs the damages one at a time. In section 3 the marginal down time distribution is obtained when there exists a repair facility for every damage (i.e. damages are repaired simultaneously).

## 1. Introduction

Correlated up time and down time models have been studied by GAVER (1972), GAVER and LUCKEW (1971), SANKARANARAYANAN and RAMANARAYANAN (1977) and RAMANARAYANAN (1977). ESARY and MARSHALL (1973) have studied a one unit system exposed to cumulative damage process where the survivor function and optimal control strategy that balances the cost of replacement with the cost of failure are of interest. ZUCKERMAN (1977) has studied the replacement policy under additive damage. SANKARANARAYANAN and RAMANARAYANAN (1977) and RAMANARAYANAN (1977) have considered the one unit system and 1-out-of-2: $F$ system models in which the system failure occurs when the cumulative damage exceeds the threshold of the device and they have obtained the correlation between the life time and the repair time by assuming that the repair time is independent of the damage magnitude. But so far no one has studied the model in which the repair time depends upon the magnitude of the damage.

Here we consider a model subjected to a cumulative damage process and the repair time depends upon the damage-magnitude. A cumulative damage process is one in which damages due to shocks are to be added to previously incurred damages until a breaking threshold of the device is reached.

In section 2, the correlation between the life time and the repair time is obtained when there is only one repair facility which repairs the damages one at a time. In section 3, the marginal down time distribution is investigated when there is a repair facility for every damage.

Let $t_0=0$ be the instant at which the device begins its function. Also at $t_0$ let the device be failure free. Let $t_1, t_2, \ldots$ be instants at which shocks occur and

[1] Department of Mathematics, Annamalai University, Annamalainagar – 608 002, S. India.

$X_1, X_2, \ldots$ are the damages caused by these shocks. Assume that the quantities $Z_n$ of the sequence $\{Z_n = t_n - t_{n-1}\}$ $(n = 1, 2, \ldots, t_0 = 0)$ are independent and identically distributed random variables with distribution function $G(x)$. Assume further that $X_1, X_2, \ldots$ are independent and identically distributed random variables independent of $\{t_n\}$. The system failure occurs when the cumulative damage exceeds a random threshold $Y$, which is distributed exponentially with parameter $\lambda$. The repair time $R_i$ of the $i$-th shock depends on the magnitude of the $i$-th damage $X_i$ and that the vector sequence $\{X_i, R_i\}_{i=1}^{\infty}$ is independent and identically distributed with a common density function $f(\cdot, \cdot)$. The repair starts immediately after the system failure and the repaired system becomes new after repair.

Let $\{X_n\}_{n=1}^{\infty}$ be a sequence of positive real valued random variables with $\{L_n(x)\}_{n=1}^{\infty}$ and $\{l_n(x)\}_{n=1}^{\infty}$ as the sequences of their distribution and density functions, then it is known that

$$\int_{S_{n-1} < y \leqq S_n} \cdots\cdots \int l_1(x_1)\, l_2(x_2) \ldots l_n(x_n)\, \mathrm{d}x_1 \ldots \mathrm{d}x_n \tag{1}$$

$$= L_1 * L_2 * \ldots * L_{n-1}(y) - L_1 * L_2 * \ldots * L_n(y)$$

where $S_n = X_1 + X_2 + \ldots + X_n$ and $*$ denotes the convolution.

## 2. Additive Repair

In this section we consider a repair facility which repairs the damages one at a time. Let $T$ and $R$ represent the time to system failure and subsequent repair time respectively. Let $N$ be the number of shocks required for the system to fail. $N$ is a random variable and hence $R = R_1 + R_2 + \ldots + R_N$ denotes the total repair time of the system when the system fails after $N$ shocks.

The joint distribution of $T$ and $R$ is given by

$$\mathsf{P}\,(T \leqq t,\, R \leqq r \mid Y = \omega)$$

$$= \sum_{n=1}^{\infty} G^{n*}(t) \left\{ \int_{s_{n-1} < \omega \leqq s_n} \cdots\cdots \int_{z=0}^{r} f_{x_1} * f_{x_2} * \ldots * f_{x_n}(z)\, \mathrm{d}z \prod_{i=1}^{n} f_1(x_i)\, \mathrm{d}x_i \right\}$$

where

(i) $f_{x_i}(\cdot) = \dfrac{f(x_i, .)}{f_1(x_i)}$, $f_1(\cdot)$ is the marginal density of the $i$-th damage

(ii) $f_{x_i} * f_{x_j}$ is the convolution of $f_{x_i}(\cdot)$ with $f_{x_j}(\cdot)$

(iii) $s_n = x_1 + x_2 + x_3 + \ldots + x_n$

(iv) $n^*$ denotes the $n$-fold convolution of the distribution function with itself.

Now

$$\mathsf{E}(\mathrm{e}^{-uT}\mathrm{e}^{-vR}) = \mathsf{E}_Y\,[\mathsf{E}_{(T,R)}(\mathrm{e}^{-uT}\mathrm{e}^{-vR} \mid Y = \omega)]$$

$$= \mathsf{E}_Y \left[ \int_0^{\infty} \int_0^{\infty} \mathrm{e}^{-uT}\mathrm{e}^{-vR} \frac{\partial^2}{\partial t \partial r} \mathsf{P}\,(T \leqq t,\, R \leqq r \mid Y = \omega) \right]$$

$$= \mathsf{E}_Y\left[\sum_{n=1}^{\infty} (\hat{G}(u))^n \int \underset{s_{n-1}<\omega \leqq s_n}{\cdots\cdots\cdots} \int \times \prod_{i=1}^{n}\left(\int_0^{\infty} e^{-v\tau_i} \frac{f(x_i, \tau_i)}{f_1(x_i)} \, d\tau_i\right) f_1(x_i)\, dx_i\right]$$

$$= \mathsf{E}_Y\left[\sum_{n=1}^{\infty} (\hat{G}(u))^n \int \underset{s_{n-1}<\omega \leqq s_n}{\cdots\cdots\cdots} \int \prod_{i=1}^{n}\left(\int_0^{\infty} e^{-v\tau_i} f(x_i, \tau_i)\, d\tau_i\right) \times dx_1 dx_2 \ldots dx_n\right]$$

$$= \lambda \int_0^{\infty} e^{-\lambda\omega}\left[\sum_{n=1}^{\infty} (\hat{G}(u))^n \int \underset{s_{n-1}<\omega \leqq s_n}{\cdots\cdots\cdots} \int \times \prod_{i=1}^{n}\left(\int_0^{\infty} e^{-v\tau_i} f(x_i, \tau_i)\, d\tau_i\right) dx_1 dx_2 \ldots dx_n\right] d\omega$$

where '$\hat{}$' denotes the LAPLACE-STIELTJES transform.

Changing the order of integration, we have

$$\mathsf{E}(e^{-uT} e^{-vR}) = \lambda \sum_{n=1}^{\infty} (\hat{G}(u))^n \left\{ \int_{\tau_1} \cdots \int_{\tau_n} e^{-v(\tau_1 + \ldots + \tau_n)} \times \left[\int_0^{\infty} e^{-\lambda\omega}\left(\int \underset{s_{n-1}<\omega \leqq s_n}{\cdots\cdots\cdots} \int \prod_{i=1}^{n} \frac{f(x_i, \tau_i)}{f_2(\tau_i)}\, dx_i\right) d\omega\right] \prod_{i=1}^{n} f_2(\tau_i)\, d\tau_i \right\} \tag{2}$$

where $f_2(\tau_i)$ is the marginal density function of $R_i$.

Let $\frac{f(\cdot, \tau_i)}{f_2(\tau_i)} = \varphi_{\tau_i}(\cdot)$, a density function and the corresponding distribution function be $\Phi_{\tau_i}(\cdot)$.

By (1) we get

$$\int_0^{\infty} e^{-\lambda\omega}\left(\int \underset{s_{n-1}<\omega \leqq s_n}{\cdots\cdots\cdots} \int \prod_{i=1}^{n} \frac{f(x_i, \tau_i)}{f_2(\tau_i)}\, dx_i\right) d\omega$$

$$= \int_0^{\infty} e^{-\lambda\omega}\left(\int \underset{s_{n-1}<\omega \leqq s_n}{\cdots\cdots\cdots} \int \prod_{i=1}^{n} \varphi_{\tau_i}(x_i)\, dx_i\right) d\omega$$

$$= \int_0^{\infty} e^{-\lambda\omega} \left(\Phi_{\tau_1} * \Phi_{\tau_2} * \ldots * \Phi_{\tau_{n-1}}(\omega) - \Phi_{\tau_1} * \Phi_{\tau_2} * \ldots * \Phi_{\tau_n}(\omega)\right) d\omega$$

$$= \frac{1}{\lambda} \prod_{i=1}^{n-1}\left(\int_0^{\infty} e^{-\lambda\omega} \frac{f(\omega, \tau_i)}{f_2(\tau_i)}\, d\omega\right) - \frac{1}{\lambda} \prod_{i=1}^{n}\left(\int_0^{\infty} e^{-\lambda\omega} \frac{f(\omega, \tau_i)}{f_2(\tau_i)}\, d\omega\right). \tag{3}$$

Substituting (3) in (2), we get

$$\mathsf{E}(e^{-uT} e^{-vR}) = \sum_{n=1}^{\infty} (\hat{G}(u))^n \left[ \int_{\tau_1} \cdots \int_{\tau_n} e^{-v(\tau_1 + \ldots + \tau_n)} \right.$$

$$\times \prod_{i=1}^{n-1} \left( \int_0^\infty e^{-\lambda\omega} f(\omega, \tau_i)\, d\omega \right) f_2(\tau_n)\, d\tau_1 \ldots d\tau_n$$

$$- \int_{\tau_1} \ldots \int_{\tau_n} e^{-v(\tau_1 + \ldots + \tau_n)} \prod_{i-1}^{n} \left( \int_0^\infty e^{-\lambda\omega} f(\omega, \tau_i)\, d\omega \right) d\tau_1 \ldots d\tau_n \Bigg]$$

$$= \sum_{n=1}^{\infty} (\hat{G}(u))^n \{[\hat{f}(\lambda, v)]^{n-1} \hat{f}(0, v) - [\hat{f}(\lambda, v)]^n\}$$

$$\mathsf{E}(e^{-uT} e^{-vR}) = \frac{\hat{G}(u)\, [\hat{f}(0, v) - \hat{f}(\lambda, v)]}{[1 - \hat{G}(u)\, \hat{f}(\lambda, v)]}. \tag{4}$$

Again from (4) we have

$$\mathsf{E}(e^{-uT}) = \frac{\hat{G}(u)\, [1 - \hat{f}(\lambda, 0)]}{[1 - \hat{G}(u)\, \hat{f}(\lambda, 0)]}$$

and

$$\mathsf{E}(e^{-vR}) = \frac{[\hat{f}(0, v) - \hat{f}(\lambda, v)]}{[1 - \hat{f}(\lambda, v)]}.$$

After differentiation, we get

$$\mathsf{E}(T) = \frac{\mathsf{E}(Z_1)}{[1 - \hat{f}(\lambda, 0)]} \tag{5}$$

and

$$\mathsf{E}(R) = \frac{\mathsf{E}(R_1)}{[1 - \hat{f}(\lambda, 0)]}, \tag{6}$$

where $\mathsf{E}(R_1)$ is the expected value of a repair time. Also from (4), we get

$$\mathsf{E}(TR) = \frac{\mathsf{E}(Z_1) \left[ \frac{\partial}{\partial \eta} \hat{f}(\lambda, v) |_{v=0} + \mathsf{E}(R_1) \right]}{[1 - \hat{f}(\lambda, 0)]^2}. \tag{7}$$

From (5), (6) and (7) we get

$$\mathsf{Cov}\,(T, R) = \frac{\mathsf{E}(Z_1) \left[ \frac{\partial}{\partial v} \hat{f}(\lambda, v) |_{v=0} \right]}{[1 - \hat{f}(\lambda, 0)]^2}.$$

From (4),

$$\mathsf{E}(e^{-sT} e^{-sR}) = \mathsf{E}(e^{-s(T+R)}) = \frac{\hat{G}(s)\, [\hat{f}(0, s) - \hat{f}(\lambda, s)]}{[1 - \hat{G}(s)\, \hat{f}(\lambda, s)]}. \tag{8}$$

Differentiating (8), we have

$$\mathsf{E}\,(T + R) = \frac{\mathsf{E}(Z_1) + \mathsf{E}(R_1)}{[1 - \hat{f}(\lambda, 0)]}. \tag{9}$$

Let $H(t)$ denote the number of breakdowns in $(0, t)$. By the BLACKWELL's Renewal Theorem and using (9) the average number of breakdowns are

$$H(t) \sim \frac{t\,[1 - \hat{f}(\lambda, 0)]}{\mathsf{E}(Z_1) + \mathsf{E}(R_1)} \quad \text{as} \quad t \to \infty .$$

### 3. Parallel Repair

In this section we assume that there exists a repair facility for every damage (i.e. damages are repaired simultaneously).

Let $\underline{R} = \text{Max}\,(R_1, R_2, \ldots, R_n)$ where $n$ is the number of damages occured to the system during the uptime $T$.

$$\mathsf{P}\,(R_i \leqq r) = \int_0^r \frac{f(x_i, \tau)}{f(x_i)}\, \mathrm{d}\tau$$

$$\mathsf{P}\,(\underline{R} \leqq r) = \prod_{i=1}^{n} \left[ \int_0^r \frac{f(x_i, \tau)}{f(x_i)}\, \mathrm{d}\tau \right]$$

$$\mathsf{P}\,(\underline{R} \leqq r, N = n \mid Y = \omega) = \int \underset{s_{n-1} < \omega \leqq s_n}{\cdots\cdots\cdots} \int \prod_{i=1}^{n} \left( \int_0^r \frac{f(x_i, \tau)}{f_1(x_i)}\, \mathrm{d}\tau \right) f_1(x_i)\, \mathrm{d}x_i$$

$$\mathsf{P}\,(\underline{R} \leqq r, N = n) = \lambda \cdot \int_0^\infty \mathrm{e}^{-\lambda\omega} \left[ \int \underset{s_{n-1} < \omega \leqq s_n}{\cdots\cdots\cdots} \int \prod_{i=1}^{n} \left( \int_0^r f(x_i, \tau)\, \mathrm{d}\tau \right) \mathrm{d}x_1 \ldots \mathrm{d}x_n \right] \mathrm{d}\omega$$

$$= \lambda \cdot \int_0^\infty \mathrm{e}^{-\lambda\omega} \left[ \int \underset{s_{n-1} < \omega \leqq s_n}{\cdots\cdots\cdots} \int \prod_{i=1}^{n} \left( \int_0^r \frac{f(x_i, \tau)}{F_2(r)}\, \mathrm{d}\tau \right) \mathrm{d}x_1 \ldots \mathrm{d}x_n (F_2(r))^n \right] \mathrm{d}\omega$$

where

$$F_2(r) = \int_0^r \int_0^\infty f(u, v)\, \mathrm{d}u \mathrm{d}v.$$

Using (1) we have

$$\mathsf{P}\,(\underline{R} \leqq r, N = n) = \left[ \int_0^r \int_0^\infty \mathrm{e}^{-\lambda u} \frac{f(u, \tau)}{F_2(r)}\, \mathrm{d}u \mathrm{d}\tau \right]^{n-1} (F_2(r))^n - \left[ \int_0^r \int_0^\infty \mathrm{e}^{-\lambda u} \frac{f(u, \tau)}{F_2(r)}\, \mathrm{d}u \mathrm{d}\tau \right]^{n} (F_2(r))^n$$

$$\mathsf{P}\,(\underline{R} \leqq r) = \sum_{n=1}^{\infty} [(K(r))^{n-1} - (K(r))^n]\, (F_2(r))^n$$

where

$$K(r) = \int_0^r \int_0^\infty \mathrm{e}^{-\lambda u} \frac{f(u, \tau)}{F_2(r)}\, \mathrm{d}u \mathrm{d}\tau$$

$$\mathsf{P}\,(\underline{R} \leqq r) = \frac{F_2(r)\,[1 - K(r)]}{[1 - K(r)\,F_2(r)]}.$$

**Acknowledgement**

We thank Prof. G. SANKARANARAYANAN for extremely helpful suggestions and guidance. We also thank the referees and the editor for their valuable suggestions and critical comments.

**References**

[1] ESARY, J. D., MARSHALL, A. W. and PROSCHAN, F.: Shock models and Wear Processes. Ann. Prob. **1**, (1973), 627–649.
[2] GAVER, D. P.: Point Process Problems in Reliability. In: Stochastic Point Processes (Ed. P. A. W. LEWIS) Wiley-Interscience, New York (1972), 774–800.
[3] GAVER, D. P. and LUCKEW: System reliability and bivariate skew distribution. Research Report, Naval Post Graduate School, Monterey, California (1971).
[4] SANKARANARAYANAN, G. and RAMANARAYANAN, R.: On correlated Life and Repair times. Math. Operationsf. u. Statistik. Ser. Optimization 8 (1977) 1, 127–136.
[5] RAMANARAYANAN, R.: General analysis of 1-out-of-2: $F$ system exposed to cumulative damage processes. Math. Operationsf. u. Statistik. Ser. Optimization 8 (1977) 2, 237–243.
[6] ZUCKERMAN, D.: Replacement models under additive damage. Nav. Res. Log. Quart. (1977) 4, 549–558.

**Zusammenfassung**

In dieser Arbeit wird die Korrelation zwischen der Lebens- und der Reparaturzeit eines Gerätes, das einem Beschädigungsprozeß ausgesetzt ist, untersucht. Die Korrelation wird in Abschnitt 2 unter der Bedingung aufgestellt, daß nicht mehrere Beschädigungen gleichzeitig repariert werden können. In Abschnitt 3 wird die randomisierte Verteilung der Stillstandszeit unter der Bedingung hergeleitet, wenn mehrere Beschädigungen gleichzeitig repariert werden können.

**Резюме**

В этой работе рассматривается корреляцию между временами жизни и ремонта прибора подвергнутого повреждению. Корреляцию устанавливается во второй части, где введена возможность ремонта, которая ремонтирует одно повреждение в некоторой времени. В третьей части устанавливаем рандомизированное распределение времени отказа, если существует возможность ремонта для любого повреждения (значит повреждения ремонтируются одновременно).

Received September 1978; revised June 1979

## Book Reviews

J. KEILSON: **MARKOV Chain Models – Rarity and Exponentiality.** Applied mathematical sciences, vol. 28. Springer-Verlag, Berlin-Heidelberg-New York 1979, XIII. 184 S., DM 24,–.

Unter den stochastischen Methoden des Operations Research nehmen die MARKOV-Ketten einen gewichtigen Platz ein. Von den überaus zahlreichen Einführungen in diesen Problemkreis hebt sich das vorliegende Buch in bemerkenswerter Weise ab. Es handelt sich hier um einen im FELLERschen Stile lehrbuchartig abgefaßten Forschungsbericht über Ideen und Resultate des Verfassers, die dieser in den letzten zehn Jahren aus der Beschäftigung mit praktischen Fragen – vornehmlich über die Zuverlässigkeit großer Systeme – gewonnen hat. Allerdings treten die ursprünglichen Fragestellungen zugunsten einer breiteren Anwendbarkeit zurück und erscheinen nun als Aussagen über das quantifizierbare Verhalten irreduzibler MARKOV-Ketten mit endlichem Zustandsraum.

Dabei ist die Verteilung der Übergangszeit des Systems von einer speziellen Anfangsverteilung in eine schwach frequentierte Teilmenge des Zustandsraumes interessant. Es werden u. a. Bedingungen studiert, unter denen die Übergangszeit näherungsweise exponentialverteilt ist.

Dieses leicht lesbare Buch ist jedem zu empfehlen, der sich von der Praktikabilität des bereitgestellten Instrumentariums zeitlich reversibler MARKOV-Ketten zur Modellierung großer Systeme bei Unsicherheit überzeugen will.

H.-J. GIRLICH

D. W. STROOCK; S. R. S. VARADHAN: **Multidimensional Diffusion Processes.** Springer-Verlag, Berlin-Heidelberg-New York 1979, 338 S., DM 70,–.

The book is the volume 233 of the famous Springer-series "Grundlehren der mathematischen Wissenschaften". Traditionally, printing and design are excellent.

The topic of the volume is the diffusion theory in $R^d$ ($=d$-dimensional EUCLIDean space). This subject fits to the general theory of MARKOV processes $X(t)$ with the state space $R^d$. In the chapters 2 and 3 the study of diffusion theory via the backward equation belonging to $X(t)$ has been shown to be a suitable approach to the subject. The major objection to this approach is that the mathematical methods used come from the theory of partial differential equations and the probabilistic input is relatively small. The chapters 4 and 5 deal with LEVY's and ITO's more probabilistic satisfactory approach which consists in considering $X(t+h) - X(t)$ for small $h$ and applying stochastic integration. With chapter 6 begins the study of diffusion theory along the lines initiated by the authors in 1968. By means of $X(t)$ they construct a stochastic process $X_\varphi(t)$ which is a martingale for all test functions $\varphi$. Using $X_\varphi(t)$ a so-called "martingale problem" is stated and shown to be the cornerstone of another theoretical treatment of diffusion. Based on this mathematical fundamental the chapters 7–12 are subjected to investigations on uniqueness problems, estimation of transition probability functions, conditions for (non-)explosion, and limit theorems.

The book is a research monograph written on a high mathematical level and having purely theoretical character. It deals with a very small branch of the theory of stochastic processes so that it will be of interest only to experts of the theory of probability and to some theoretical physicists.

F. Beichelt

G. Färber: **Prozeßrechentechnik.** Springer-Verlag, Berlin-Heidelberg-New York 1979, 208 S., 98 Abb., DM 48,–.

Die von H. Marko herausgegebene Reihe zur Nachrichtentechnik kann für sich in Anspruch nehmen, ein sehr weites und hochaktuelles Wissensgebiet komplex zu erfassen. Mit dem Band 5 wird die Prozeßrechentechnik vorgestellt, deren Bedeutung in den vergangenen Jahren enorm angewachsen ist. Mit den drei Teilen des Buches: *Prozeßautomatisierung*, *Prozeßrechner-Hardware* und *Prozeßrechner-Software* werden moderne Grundbegriffe wie "online", "ofline", "closed loop" Datentypen. Interface, Architektur, "realtime", Adressierungsarten, Operationssystem, Datenformate, Programmiersprachen, Befehlsarten, Übersetzer, A/D-, D/A-Signalwandler, Mikroprogrammierung Codes PDV-Bus u. a. voraussetzungsfrei behandelt, so daß ihr direktes Verständnis ungeteilt möglich ist.

Die enge Verbindung von Prozeßrechentechnik, Prozeßanalyse und Prozeßoptimierung erfordern übergreifende Kenntnisse, wie sie hier effektiv angeboten werden.

Das Buch kann schnell gelesen werden. Es gewährleistet die Kenntnisvermittlung der wichtigsten hochaktuellen Begriffe und Auffassungen der Prozeßrechentechnik, wie sie sich auf der Basis der Mikroelektronik herausgebildet hat und in zahlreichen Gebieten der industriellen Produktion, der Energietechnik, Verkehrstechnik, in der Laborautomatisierung und der Medizin durchsetzt.

Das Buch ist allen denjenigen zu empfehlen, die einen gesicherten minimalen Einblick in dieses aktuelle und zukunftsreiche Gebiet gewinnen wollen.

M. Roth

M. Zeleny (Hrsg.): **Multiple Criteria Decision Making.** Kyoto 1975. Lecture Notes in Economics and Mathematical Systems. Vol. 123. Springer-Verlag, Berlin-Heidelberg-New York 1976, 345 S., DM 32,–.

Entscheidungsprobleme mit mehrfacher Zielsetzung insbesondere Vektoroptimierungsaufgaben (auch Polyoptimierungsaufgaben genannt) spielen bei der Lösung ökonomischer, technischer und naturwissenschaftlicher Problemstellungen eine immer größere Rolle. Die Session „Multiple Criteria Decision Making“ des XXII. Internationalen Meetings des TIMS (The Institute of Management Sciences) befaßte sich mit dieser Thematik. Im vorliegenden Band sind fünfzehn Vorträge publiziert, die auf dieser Konferenz gehalten wurden. Es werden verschiedene Lösungskonzepte für derartige Entscheidungsprobleme diskutiert (z. B. das Konzept der Nutzenfunktion auf der einen Seite und das einer geeigneten Kompromißmenge auf der anderen Seite). Algorithmen zur Bestimmung von „Lösungen“ wurden vorgeschlagen, z. B. eine Modifikation der Simplexmethode zur Bestimmung „nichtdominierender“ Extremalpunkte linearer Vektoroptimierungsprobleme einschließlich eines Fortranprogramms.

Die praktische Bedeutung wird z. B. durch die Anwendung auf ein Wasser-Resourcen-Problem unterstrichen.

Es werden auch Probleme des menschlichen Urteilsvermögens vom Standpunkt des Psychologen aus diskutiert.

Eine Bibliographie mit ungefähr fünfhundert Literaturstellen gibt eine Übersicht über die Publikationen bis zum Jahre 1975, in der allerdings viele Arbeiten von sowjetischen Spezialisten nicht enthalten sind.

J. Guddat

H. P. KÜNZI; W. KRELLE; R. v. RANDOW: **Nichtlineare Programmierung.** Zweite, neubearbeitete und erweiterte Auflage. Springer-Verlag, Berlin-Heidelberg-New York 1979, 262 S., 36 Abb., DM 49,–.

Eines der ersten bedeutenden Lehrbücher auf dem Gebiet der nichtlinearen Optimierung ist das 1962 als erster Band der Reihe Ökonomie und Unternehmensforschung erschienene Buch „Nichtlineare Programmierung" von H. P. KÜNZI und W. KRELLE. Der Springer-Verlag hat nun in der Reihe Hochschultext eine zweite, neubearbeitete und erweiterte Auflage herausgegeben, in der einerseits Bewährtes aus der ersten Auflage übernommen wurde und andererseits zusätzlich aufgenommene Kapitel der stürmischen Entwicklung des Gegenstandes in den letzten Jahren Rechnung tragen. Neben dem neuen Mitautor wirkte auch W. OETTLI an dieser Ausgabe mit.

Das Buch ist in drei Teile untergliedert, wobei die beiden ersten im wesentlichen aus der ersten Auflage übernommen wurden. Im ersten Teil werden einige allgemeine mathematische Hilfsmittel bereitgestellt und erfolgt eine Einführung in die Theorie der linearen und nichtlinearen Optimierung. Das Hauptanliegen des Buches besteht jedoch in der Darstellung und Erläuterung von Lösungsverfahren der nichtlinearen Optimierung. Die Ausführungen zur Theorie sind dem untergeordnet.

Der zweite Teil des Buches enthält ausgewählte Verfahren zur Lösung quadratischer Optimierungsprobleme. Es handelt sich dabei um die Verfahren von HILDRETH und d'ESOPO, BEALE, WOLFE, BARANKIN und DORFMAN, FRANK und WOLFE, HESTENES und STIEFEL, ROSEN sowie von ZOUTENDIJK. Sämtliche dieser Verfahren werden an durchgerechneten Zahlenbeispielen gut illustriert. Gegenüber der ersten Auflage nicht mehr aufgenommen wurden die Methoden von THEIL und VAN DE PANNE, FRISCH und von HOUTHAKKER, die sich in der Entwicklung des Gebiets als nicht so zweckmäßig herausgestellt haben.

Völlig neu ist der dritte Teil des Buches, der sich mit Verfahren zur Lösung allgemeiner nichtlinearer Optimierungsprobleme beschäftigt. Im ersten Kapitel dieses Teils (im 13. Kapitel insgesamt) wird die Einführung in die nichtlineare Optimierung weitergeführt. Das 14. Kapitel enthält Verfahren zur eindimensionalen Optimierung (FIBONACCI-Verfahren, Verfahren des Goldenen Schnittes, Verfahren von POWELL und von SWANN). Derartige eindimensionale Probleme treten bekanntlich bei der Durchführung vieler Methoden der nichtlinearen Optimierung als Hilfsprobleme auf. Im 15. Kapitel sind sieben Verfahren für Probleme ohne Nebenbedingungen zusammengefaßt, nämlich das Verfahren des steilsten Abstiegs, die Verfahren von NEWTON, von DAVIDON, FLETCHER und POWELL, von BROYDEN, FLETCHER, GOLDFARB und SHANNO und von FLETCHER und REEVES sowie das ableitungsfreie Verfahren von POWELL und das Verfahren der Rang-1-Korrektur. Die letzten sieben Kapitel enthalten sieben Verfahrensklassen für Aufgaben mit Restriktionen. Behandelt werden das Verfahren von TOPKIS und VEINOTT, die Methode der reduzierten Gradienten, Schnittebenenverfahren (KELLEY, KLEIBOHM und VEINOTT), innere und äußere Straffunktionsverfahren und die Zentrenmethode von HUARD.

Insgesamt kann man einschätzen, daß dieses Buch in seiner spezifischen Weise eine gute Einführung in den aktuellen Entwicklungsstand der Verfahren der nichtlinearen Optimierung gibt. Sein besonderer Vorzug besteht in der Vielfalt der betrachteten Methoden zur quadratischen und allgemeinen nichtlinearen Optimierung. Die Darstellung ist verständlich und mathematisch korrekt. Alle Verfahren werden gut motiviert, begründet und illustriert. Auf einige Vor- und Nachteile einzelner Methoden, auf noch offene Fragen sowie auf weiterführende Literatur wird an vielen Stellen hingewiesen. Allen, die sich in Lehre, Forschung oder Anwendung mit Fragen der mathematischen Optimierung beschäftigen, kann dieses Buch empfohlen werden.

KLAUS TAMMER

W. Oberhofer: **Lineare Algebra für Wirtschaftswissenschaftler.** R. Oldenbourg Verlag München 1978, 213 S., DM 39,50.

Der Verfasser ist ord. Professor für Ökonometrie an der Universität Regensburg. Nach seinen Angaben ist das Buch aus einer einsemestrigen Vorlesung für Studenten der Wirtschaftswissenschaften hervorgegangen. Nach einführenden Bemerkungen zum Meßproblem, das auch in den weiteren Ausführungen berücksichtigt wird, werden im Teil 1 Vektoren, Matrizen und lineare Gleichungssysteme behandelt. In diesem Teil wird jeweils von einfachen ökonomischen Beispielen ausgegangen und an ihnen gezeigt, warum und wie Begriffe, Relationen und Operationen zu wählen sind. Dabei wird für Wirtschaftswissenschaftler sehr sinnvoll der Vektor als $n$-Tupel eingeführt. Anschließend erfolgt die geometrische Darstellung als Pfeil in der Ebene und im Raum, wobei das kartesische Koordinatensystem und das Polarkoordinatensystem genutzt werden. Auch der Matrizenbegriff, Matrizenrelationen und -operationen werden sehr anschaulich an Hand von ökonomischen Beispielen eingeführt. Die vom Autor getrennte Behandlung von Vektoren und Matrizen erleichtert zwar eventuell das Verständnis, führt aber durch die Dopplung zu unnötiger Ausweitung.

Auch die linearen Gleichungssysteme werden sehr anschaulich eingeführt. Zur Lösung wird der Gausssche Algorithmus genutzt.

Im Anschluß an die Behandlung der linearen Gleichungssysteme werden dann Linearkombinationen, die lineare Abhängigkeit von Vektoren und der Rang einer Matrix eingeführt. Mit dem Rangbegriff werden dann Aussagen über die Lösbarkeit und die Lösungsmenge linearer Gleichungssysteme gemacht. Dieses Vorgehen ist methodisch günstig, weil bei der Überprüfung der linearen Abhängigkeit von Vektoren und bei der Rangbestimmung die Kenntnisse aus der Lösung linearer Gleichungssysteme genutzt werden können.

Im zweiten Teil des Buches werden die Matrizeninversion, Determinanten und Elemente der analytischen Geometrie vorgestellt. Während im Teil 1 des Buches ein starker ökonomischer Bezug gefunden wurde, entfällt jetzt dieser Bezug völlig.

Die Matrizeninversion wird über die Lösung von linearen Gleichungssystemen durchgeführt. Als Ausgangspunkt für die Einführung von Determinanten wird das Problem der Flächen- bzw. Volumenberechnung gewählt. Für die konkrete Berechnung der Determinante wird der Gausssche Algorithmus genutzt, indem die zugehörige Matrize in eine Dreiecksmatrix umgeformt wird. Als Anwendung der Determinanten werden die Matrizeninversion und die Cramersche Regel gebracht.

In der Analytischen Geometrie wird zunächst die Gerade behandelt. Sie wird sowohl in allgemeiner Form und kartesischer Normalform als auch mit Hilfe von Vektoren in der Parameterdarstellung und der Hesseschen Normalform vorgeführt. Als Anwendung wird die Lösung linearer Gleichungssysteme gebracht. Anschließend wird die Darstellung der Ebene und der Geraden im Raum vorgenommen. Es erfolgt auch ein kurzer Hinweis auf die Geometrie im $R^n$ für $n > 3$. Einen breiteren Raum nehmen die Koordinatentransformation und die lineare Transformation ein.

Im dritten Teil des Buches werden das Eigenwertproblem und quadratische Formen behandelt. Bei der Einführung des Eigenwertproblems wird wieder von einem ökonomischen Problem ausgegangen. Die anschließende theoretische Darstellung ist für Wirtschaftswissenschaftler sehr anspruchsvoll, weil sie, obwohl auf komplexe Zahlen verzichtet wird, doch einen hohen Grad an Abstraktion verlangt. Der Hinweis auf Verflechtungsmodelle (Input-Output-Analyse) bei der Vorführung der Neumannschen Reihe entschärft diesen Sachverhalt kaum. Zum Abschluß werden quadratische Formen mit Hilfe der Eigenwerttheorie behandelt.

Das Verständnis des Lesers wird durch viele Beispiele, überwiegend ökonomischer Natur, durch Abbildungen, besondere Kennzeichnung von Definitionen, Sätzen und Beweisen und Hervorhebung wichtiger Begriffe gefördert. Neben den zahlreichen Beispielen gibt es zu den einzelnen Abschnitten Übungsaufgaben, die sinnvoll durch einen Lösungsteil am Ende des Buches ergänzt werden. Die Übungsaufgaben stellen recht unterschiedliche Anforderungen an den Leser, der Autor führt aus diesem Grunde für Aufgaben weiterführender Natur eine besondere Kennzeichnung ein. Für einige Übungsaufgaben, speziell zur Problematik der Verflechtung, wäre eine Berücksichtigung im laufenden Text ein Gewinn, weil ein stärkerer Praxisbezug an manchen Stellen die Anforderungen an die Abstraktion gemildert und Überforderungen in den Übungen abgebaut hätte. In diesem Zusammenhang soll nochmals festgestellt werden, daß die Anforderungen an den Leser recht unterschiedlich sind, fortschreitend mit der Seitenzahl wachsen die Anforderungen. Dies ist einerseits natürlich, wird aber durch den Autor zusätzlich dadurch beeinflußt, daß sich die Vorgehensweise vom fast plaudernden Ton bis zur anspruchsvolleren abstrakten Darstellung völlig ändert.

Ein Inhaltsverzeichnis am Anfang und ein Sachwortverzeichnis am Ende des Buches erleichtern den Umgang mit dem Buch. An manchen Stellen des Buches überrascht den Fachmann die Ausdrucksweise, wenn Zeilen eines Gleichungssystems oder Elemente „verschwinden“, wenn Gleichungen „halbiert“ und „erweitert“ werden, das „$t$-fache“ einer Gleichung auftritt, wenn Gleichheit, Summe und Differenz von Vektoren als „Beziehungen“ eingeführt werden.

Der Titel des Buches „Lineare Algebra für Wirtschaftswissenschaftler“ kennzeichnet bereits den vorgesehenen Leserkreis, wobei sowohl Studierende als auch in der Praxis oder in der Wissenschaft tätige Wirtschaftswissenschaftler angesprochen werden. Aber auch der Lehrende der Linearen Algebra kann einige methodische Anregungen entnehmen.

H. Körth

T. Gal: **Postoptimal Analysis, Parametric Programming, and Related Topics.** McGraw-Hill New York 1979, 380 S., DM 104,–.

Das vorliegende Buch stellt eine überarbeitete englische Übersetzung seiner 1973 erschienenen „Betriebliche Entscheidungsprobleme, Sensitivitätsanalyse und parametrische Programmierung“ dar. Darin ist das frühere erste Kapitel über die Modellierung betrieblicher Systeme als „out-of-date“ gestrichen worden, so daß das Buch nunmehr einen ausschließlich mathematischen Inhalt hat. Es beginnt jetzt sofort mit einem resümierenden Kapitel über die Grundlagen der linearen Optimierung, wobei aber Vorkenntnisse in linearer Optimierung und in linearer Algebra beim Leser vorausgesetzt werden. Neu hinzugekommen sind aus dem unmittelbaren Arbeitsgebiet des Autors eine graphentheoretische Analyse der Entartung und ihr Zusammenhang mit der Redundanz. Das nächste Kapitel behandelt suboptimale Lösungen in Abhängigkeit von den Nichtbasisvariablen der optimalen Lösung und geht dann ausführlich auf den Begriff der Redundanz ein. Es wird auch ein Algorithmus zur Bestimmung redundanter Restriktionen angegeben. Das dritte Kapitel befaßt sich mit der Sensitivitätsanalyse bezüglich der rechten Seite. Dabei werden lineare Variationen der rechten Seite in einem und mehreren Parametern und die entsprechenden kritischen Bereiche (Stabilitätsbereiche) betrachtet, innerhalb welcher eine optimale Basislösung optimal bleibt. Zur Approximation der nur schwer berechenbaren kritischen Bereiche durch eine Teilmenge wird die konvexe Hülle der kritischen Segmente der Einzelparameter herangezogen. Kapitel 4 stellt vom Umfang und Inhalt her ein Hauptkapitel des Buches dar; in diesem wird die lineare parametrische Optimierung zunächst auch bezüglich der rechten Seite ausführlich dargestellt. Der Autor beginnt mit bekannten Ergebnissen und Methoden für das einparametrische Problem und geht dann zu mehrparametrischen Problemen über, wobei der homogene Fall gesondert behandelt wird. Um die beträchtlichen Schwierigkeiten bei der vollständigen Behandlung des mehrparametrischen Problems, d. h. bei der Bestimmung aller kritischen Bereiche zu überwinden, entwickelt der Autor einen Lösungsalgorithmus, der auf dem Simplexalgorithmus und auf der Verwendung graphentheoretischer Elemente beruht. Eine Einschätzung seiner praktischen Leistungsfähigkeit wird jedoch nicht gegeben. In den nächsten beiden Kapiteln wird in entsprechender Weise die Sensitivitätsanalyse und parametrische Optimierung bezüglich der Zielfunktion behandelt, während in Kapitel 7 die gleichzeitige Abhängigkeit der rechten Seite und der Zielfunktion von denselben Parametern untersucht wird. In Kapitel 8 macht der Autor einige Ausführungen zur Parameterabhängigkeit der Koeffizientenmatrix. Wie bekannt, führt aber dieser Fall völlig aus den bisherigen Strukturen heraus – die kritischen Bereiche brauchen nicht mehr konvex und die Optimalwertfunktion nicht mehr stetig zu sein – so daß es hier keine abgerundete Theorie und keine befriedigende Lösungsmethoden gibt. Neu aufgenommen wurde in die vorliegende englische Ausgabe das Kapitel 9 über „Multicriteria linear programming", in welchem Vektoroptimierungsprobleme und ihr Zusammenhang mit der parametrischen Optimierung behandelt und ein Verfahren zur Bestimmung aller effizienten Lösungen angegeben werden. Abschließend wird in Kapitel 10 noch auf die Variation von Basisvariablen und auf gleichzeitige Variation von Matrix- und Zielfunktionskoeffizienten sowie auf Inkonsistenz, Degeneration und Redundanz eingegangen, sowie einige Ausführungen zum parametrischen Transportproblem gemacht. Das Buch enthält eine umfassende Bibliographie mit einem ergänzenden Nachtrag bis etwa 1978. Es fehlt darin die Monographie über parametrische Optimierung von F. Nozicka et al., Berlin 1974.

Das vorliegende Buch gibt einen klaren und übersichtlichen Einblick in Theorie und Lösungsalgorithmen der linearen parametrischen Optimierung. Es leitet seine Motivation zwar von einer ökonomischen Problematik her, hat sonst aber rein mathematischen Charakter, wenn man von einigen ökonomischen Termini absieht wie z. B. technologische Koeffizienten oder Kostenkoeffizienten (in einer zu maximierenden (!) Zielfunktion). In didaktischer Hinsicht weist das Buch eine Besonderheit auf. Alle Kapitel bestehen aus zwei Teilen; im jeweils ersten Teil werden die zu behandelnden Probleme anhand einfacher formal-mathematischer Beispiele diskutiert und erst im zweiten, meist kürzeren Teil folgt ihre allgemeine mathematische Behandlung. Wenn auch diese Vorgehensweise vielleicht dem anwendungsorientierten Leser entgegenkommen soll, so ist sie doch problematisch, da durch die einführenden Beispiele auch falsche Vorstellungen geweckt werden können. So tritt z. B. in Beispiel 1–4 nur eine künstliche Variable auf, was fälschlicherweise für typisch gehalten werden könnte, und die Entartung wird in Beispiel 1–6 durch Anbringung einer redundanten Restriktion eingeführt, was ebenfalls auf einen falschen Zusammenhang mit der Redundanz könnte schließen lassen. Es sollte dem Leser deshalb dringend geraten werden, auch die zweiten Teile zu lesen. Es gibt in der Mathematik keinen Königsweg.

J. Focke

## Informations

### Bericht über das XXIV. Internationale Wissenschaftliche Kolloquium (Vortragsreihen B1 bis B3) der Technischen Hochschule Ilmenau

vom 22. bis 26. 10. 1979 in Ilmenau – DDR

Die von den Sektionen Mathematik, Rechentechnik und ökonomische Kybernetik (Vortragsreihen B1 bis B4) und Technische und Biomedizinische Kybernetik (Vortragsreihen A1 bis A4) der TH Ilmenau getragene internationale Tagung wurde von 1166 Wissenschaftlern besucht, und es wurden insgesamt 275 Vorträge gehalten. In diesem Bericht wird ein Überblick über die Thematik und die gehaltenen Vorträge gegeben, die die mathematischen Disziplinen betreffen. Diese Tagung ist für die auf den Gebieten der mathematischen Optimierung einschließlich der Theorie der optimalen Steuerung, der Graphen- und Netzwerktheorie und der Simulation (auf Rechnern) arbeitenden Wissenschaftlern zu einem wichtigen Treffpunkt für den Erfahrungsaustausch, die Vorstellung neuer Ergebnisse und das Setzen neuer Impulse zur Strategie der Forschungsarbeit in den nächsten Jahren geworden.

In der Reihe B1 „Mathematische Optimierungstheorie und ihre Anwendung" (Reihenverantwortlicher: Prof. Dr. K.-H. Elster) wurden die Teildisziplinen behandelt:

- Theorie der mathematischen Optimierung (insbesondere nichtlineare, stochastische, parametrische, ganzzahlige und dynamische Optimierung);
- Anwendungen der mathematischen Optimierung auf technische, ökonomische und andere Probleme;
- Spezielle Optimierungsprobleme;
- Lösungsverfahren der mathematischen Optimierung;
- Theorie der optimalen Steuerung.

Die Vorträge und Diskussionen gestatten folgendes Resumee:

- Neben einer Vorstellung neuer Ergebnisse fand ein reger Erfahrungsaustausch über Tendenzen der wissenschaftlichen Entwicklung auf dem Gebiet der mathematischen Optimierung und der Theorie der optimalen Steuerung statt.
- Numerische Erfahrungen mit Optimierungsmodellen wurden ausgetauscht, und es erfolgte eine kritische Analyse ihrer Leistungsfähigkeit (Anforderungen an vorzugebende technische und ökonomische Daten, Rechenaufwand, Integration der Optimierungsergebnisse in der Planung, Ökonomie und Technik).
- Von besonderer Bedeutung war die Teilnahme einer Delegation der Akademie der Wissenschaften der UdSSR unter der Leitung von Prof. Worobjoff (Leningrad). Mit dieser Delegation wurden mehrere Disziplinen der mathematischen Optimierung, einschließlich ihrer Beziehungen zur Kybernetik, Numerik und Graphentheorie repräsentiert.
- Ein Schwerpunkt der Tagung waren die Vorträge zur Vektoroptimierung, die sich zur Zeit in einer außerordentlich raschen Entwicklung befindet.

In der Reihe B2 „Theorie der Graphen und Netzwerke und deren Anwendungen" (Reihenverantwortlicher: Prof. Dr. H. Sachs) standen Fragen der Strukturtheorie sowie Anwendungen graphen- und netzwerktheoretischer Methoden in der physikalisch-technischen und technisch-ökonomischen Praxis im Vordergrund. Schwerpunkte waren:

- Strukturtheorie endlicher Graphen;
- Anwendungen graphentheoretischer Methoden und Algorithmen zur Lösung von Optimierungsaufgaben;
- Netzplantechnik.

Besonders deutlich waren qualitative und quantitative Erfolge bei Untersuchungen auf dem Gebiet der Theorie der Effektivität von Algorithmen (NP-Vollständigkeit gewisser Problemklassen usw.). Es wurden Möglichkeiten aufgezeigt, die Matroidtheorie zur Berechnung (Analyse, später Synthese) aktiver elektrischer Netze (mikroelektronische Systeme mit gesteuerten Quellen) einzusetzen. Hervorzuheben sind weiter neue Erkenntnisse über Strukturaussagen von Graphen (Spektrum des Graphen, Existenz spezieller Kreise in Graphen und algebraische Methoden in der Graphentheorie), Aussagen über die Effektivität neu vorgestellter Algorithmen in der Anwendung der Graphentheorie und Färbungsprobleme.

Die Reihe B3 „Simulation auf Rechnern" (Reihenverantwortlicher: Prof. Dr. R. Schönefeld) beschäftigte sich mit Ergebnissen auf dem Gebiet der Systemsimulation auf Hybrid- und Digitalrechnern.
Im Mittelpunkt standen hierbei:
– Grundlagen der Simulation;
– Simulationssoftware und Anwendungen in verschiedenen Wissenschaftsgebieten.
Neben der Vorstellung und Diskussion neuer theoretischer Ergebnisse und Probleme sind auch Beiträge mit der Darlegung von Aufgabenstellungen, die aus der Praxis stammen und zur weiteren theoretischen Bearbeitung Anlaß geben, hervorzuheben. Bemerkenswerte Resultate wurden in Vorträgen über rechentechnische Methoden zur Lösung von partiellen Differentialgleichungen, Genauigkeitsuntersuchungen eines Multiprozessorsystems für Randwertaufgaben und über Möglichkeiten zur hybriden und digitalen Simulation mathematischer, ökonomischer und technischer Prozesse vorgestellt.

Weiter wurden eine optimale Vermittlungstechnik für den Informationsfluß in Datenfernverarbeitungssystemen und ein System zur Modellierung von kleinen Dreizustandsnetzwerken zur Diskussion gestellt.

B. Marx

**Bericht über die 4. Fachtagung „Rechnergestützte Optimierung"**
am 8. und 9. 5. 1980 in Karl-Marx-Stadt, DDR

Die Arbeitsgemeinschaft „Rechnergestützte Optimierung von Konstruktionen" der Kammer der Technik, Bezirksverband Karl-Marx-Stadt, führte in Verbindung mit der Technischen Hochschule Karl-Marx-Stadt, Sektion Maschinen-Bauelemente, Wissenschaftsbereich Konstruktionen, der Betriebssektion der Kammer der Technik des VEB Werkzeugmaschinen-Kombinat „Fritz Heckert" Karl-Marx-Stadt (Stammbetrieb) und der Mathematischen Gesellschaft der DDR, Bezirkssektion Karl-Marx-Stadt am 8. und 9. 5. 1980 ihre 4. Fachtagung zur Thematik „Rechnergestützte Optimierung" durch. An dieser Tagung nahmen Mitarbeiter aus volkseigenen Betrieben des Maschinen- und Apparatebaus, des Schiffsbaus, der chemischen Industrie und des Bauwesens, Vertreter der entsprechenden technischen und mathematischen Forschungsrichtungen in Universitäten und Hochschulen der DDR sowie der Akademie der Wissenschaften und der Bauakademie der DDR teil. Als Gäste konnten auch einige Mitarbeiter entsprechender Einrichtungen aus dem sozialistischen Ausland begrüßt werden.

In der Plenarsitzung am ersten Tag sprach nach der Begrüßung durch Vertreter der veranstaltenden Organisation J. Müller (Akademie der Wissenschaften der DDR) zur „Rationalisierung des konstruktiven Entwicklungsprozesses – Stand und Perspektiven". Danach berichtete H. Aurich (TH Karl-Marx-Stadt) über „Bildschirmunterstütztes Konstruieren im Maschinenbau", gefolgt von F. Rudolf (TH Leipzig), der über „Rechnergestützte Optimierung von Tragwerken im Bauwesen" vortrug. Damit wurde der Stand der Anwendung von Optimierungstrategien in zwei der wichtigsten technischen Lehr- und Forschungsgebiete in Beispielen aufgezeigt.

Die Arbeit der Tagung wurde dann mit 33 Kurzvorträgen in vier Sektionen (Maschinen- und Gerätebau, Verfahren und Prozesse, Bauwesen, Mathematik) fortgesetzt, wobei inhaltlich mitunter ein Übergreifen zwischen diesen Überschriften unvermeidbar war. Aus der Vielzahl der Themen seien, auch aus Platzgründen, nur die folgenden nach Auffassung des Berichterstatters im Rahmen dieser Zeitschrift besonders relevanten Ausführungen mitgeteilt, ohne dadurch eine Wertung vorzunehmen:

J. Matyàš (TESLA Pardubice/ČSSR) stellte ein „Optimierungsverfahren mit stochastischer Strategie" vor. Mit Fragen der Realisierung der Optimierung mit mehrfacher Zielsetzung beschäftigten sich J. Ester (TH Karl-Marx-Stadt) mit „Eine iterative Prozedur als Hilfsmittel bei der Lösung stabiler Polyoptimierungsaufgaben" und D. Franz (TH Karl-Marx-Stadt) mit „Aufbau des Programmsystems Polyoptimierung". K.-D. Weber (TH Ilmenau) erläuterte „Algorithmus und Programm zur nichtlinearen Ausgleichsrechnung". G. Langensiepen (TU Dresden) stellte in „Ein Programmpaket zur nichtlinearen Optimierung für ESER-Anlagen" die Weiterentwicklung der für die Anlage BESM-6 geschaffenen Software vor. C. Richter (TU Dresden)/S. Bermann (VEB Kombinat Mikroelektronik Erfurt) erläuterten durch einen Bericht „Einsatz von Optimierungsprogrammen in der Mikroelektronik-Technologie" die Anwendung des Programmpakets in der technischen Praxis. Dem Einsatz von Kleinrechnern war ein Vortrag von G. Donath (PH Halle) „Ein Programmpaket zur nichtlinearen Optimierung für das KRS 4201" gewidmet.

Die Tagung klang am zweiten Tag mit zwei Plenarvorträgen aus: K.-H. Elster (TH Ilmenau) sprach über „Einige neuere Entwicklungen bei dem Lösungsverfahren für nichtlineare Optimierungsprobleme" wobei er, ausgehend von bekannten Strafverfahren, das Konzept der Primal-Dual-Methode vorstellte. A. Dudszus (Wilhelm-Pieck-Universität Rostock) vermittelte sehr anschaulich einen Eindruck von den Problemen der „Entwurfs- und Konstruktionsoptimierung in der Schiffstechnik".

G. Donath

## Announcement

**IX. Internationaler Kongreß über Anwendungen der Mathematik in den Ingenieurwissenschaften (IX. IKM)**
vom 28. Juni bis 4. Juli 1981 in Weimar
Hochschule für Architektur und Bauwesen Weimar

*Thema:* Anwendung mathematischer Methoden und der EDV zur Intensivierung der Bauproduktion

*Sektionen und Themenschwerpunkte:*

*Sektion 1: Untersuchung von Tragwerken*
– Dynamisches Verhalten und Stabilitätsverhalten von Flächentragwerken
– Rißbildung und Grenztragfähigkeit von Betontragwerken
– Bauwerksverhalten unter extremen Bedingungen

*Sektion 2: Nutzung mathematischer Methoden in der Betriebswirtschaft*
– Mathematische Methoden bei der Produktionsvorbereitung und -durchführung
– Leitung und Planung des betrieblichen Reproduktionsprozesses
– Mathematische und kybernetische Aspekte modell- und rechnergestützter Leitungssysteme

*Sektion 3: Automatisierte Projektierung*
– Automatengestützte Optimierung von Projektierungslösungen
– Mensch-Maschine-Kommunikation
– Gestaltung und Entwicklung automatisierter Projektierungssysteme

*Sektion 4: Anwendungen der numerischen Mathematik*
– Numerische Methoden für nichtlineare Probleme
– Besonderheiten bei der Lösung nichtkorrekter Aufgaben
– Methoden der finiten Elemente

*Sektion 5: Methoden und Modelle der Stochastik*
– Parameterschätzung stochastischer Modelle
– Versuchsplanung
– Modelle der Bedienung, Lagerhaltung und Zuverlässigkeit.

*Sektion 6: Methoden und Modelle der Optimierung*
– Effektive Algorithmen zur Behandlung großer Probleme
– Untersuchungen zur Koppelung und Zerlegung von Modellen
– Ablaufplanung und Steuerungsprobleme

*Auskünfte:* Hochschule für Architektur und Bauwesen Weimar, Kongreßbüro IX. IKM
DDR-5300 Weimar, Karl-Marx-Platz 2

Bezugsmöglichkeiten der Zeitschrift „Mathematische Operationsforschung und Statistik – Series Optimization“:
Bestellungen sind zu richten

– *in der DDR* an den Postzeitungsvertrieb, an eine Buchhandlung oder an den AKADEMIE-VERLAG, DDR - 1080 Berlin, Leipziger Str. 3–4
– *im sozialistischen Ausland* an eine Buchhandlung für fremdsprachige Literatur oder an den zuständigen Postzeitungsvertrieb
– *in der BRD und Westberlin* an eine Buchhandlung oder an die Auslieferungsstelle KUNST UND WISSEN, Erich Bieber, D - 7000 Stuttgart 1, Wilhelmstr. 4–6
– *in Österreich* an den Globus-Buchvertrieb A - 1201 Wien, Höchstädtplatz 3
– *in den übrigen westeuropäischen Ländern* an eine Buchhandlung oder an die Auslieferungsstelle KUNST UND WISSEN, Erich Bieber GmbH, CH - 8008 Zürich/Schweiz, Dufourstraße 51
– *im übrigen Ausland* an den Internationalen Buch- und Zeitschriftenhandel; den Buchexport, Volkseigener Außenhandelsbetrieb der Deutschen Demokratischen Republik, DDR - 7010 Leipzig, Postfach 160, oder an den AKADEMIE-VERLAG DDR - 1080 Berlin, Leipziger Straße 3–4

Zeitschrift „Mathematische Operationsforschung und Statistik – Series Optimization“

Herausgeber: Akademie-Verlag, DDR - 1080 Berlin, Leipziger Straße 3–4; Fernruf 2 23 62 21 oder 2 23 62 29; Telex-Nr. 11 44 20; Bank: Staatsbank der DDR, Berlin, Kto.-Nr. 6836-26-27012

Chefredakteur: Prof. Dr. Karl-Heinz Elster
Verantwortlicher Redakteur: Dr. Rüdiger Reinhardt

Anschrift der Redaktion: Sektion Mathematik, Rechentechnik und ökonomische Kybernetik der TH Ilmenau, DDR - 6300 Ilmenau, Postfach 327

Veröffentlicht unter der Lizenznummer 1586 des Presseamtes beim Vorsitzenden des Ministerrates der Deutschen Demokratischen Republik

Gesamtherstellung: VEB Druckerei „Gottfried Wilhelm Leibniz“, DDR - 4450 Gräfenhainichen

Erscheinungsweise: Die Zeitschrift „Mathematische Operationsforschung und Statistik – Series Optimization“ erscheint jährlich in einem Band mit 4 Heften. Bezugspreis je Band 120,– M zuzüglich Versandspesen (Preis für die DDR 80,– M); Preis je Heft 30,– M (Preis für die DDR 20,– M)

Bestellnummer dieses Heftes: 1093/11/4



Printed in the German Democratic Republic.

AN (EDV) 40234

## Contents



### Book Reviews



### Informations



### Announcement